PENGUIN BOOKS

MAKING THE FUTURE

Noam Chomsky is the author of numerous bestselling and influential political books, including *Hegemony or Survival*, *Failed States*, *Interventions*, *What We Say Goes*, *Hopes and Prospects* and *Gaza in Crisis*, all of which are published by Hamish Hamilton/Penguin. He is an Institute Professor (Emeritus) in the Department of Linguistics and Philosophy at MIT, and is widely credited with having revolutionized modern linguistics. He lives in Lexington, Massachusetts.

Making the Future

*Occupations, Interventions,
Empire and Resistance*

Noam Chomsky

PENGUIN BOOKS

PENGUIN BOOKS

Published by the Penguin Group
Penguin Books Ltd, 80 Strand, London WC2R 0RL, England
Penguin Group (USA) Inc., 375 Hudson Street, New York, New York 10014, USA
Penguin Group (Canada), 90 Eglinton Avenue East, Suite 700, Toronto, Ontario, Canada M4P 2Y3
(a division of Pearson Penguin Canada Inc.)
Penguin Ireland, 25 St Stephen's Green, Dublin 2, Ireland (a division of Penguin Books Ltd)
Penguin Group (Australia), 707 Collins Street, Melbourne, Victoria 3008, Australia
(a division of Pearson Australia Group Pty Ltd)
Penguin Books India Pvt Ltd, 11 Community Centre, Panchsheel Park, New Delhi – 110 017, India
Penguin Group (NZ), 67 Apollo Drive, Rosedale, Auckland 0632, New Zealand
(a division of Pearson New Zealand Ltd)
Penguin Books (South Africa) (Pty) Ltd, Block D, Rosebank Office Park,
181 Jan Smuts Avenue, Parktown North, Gauteng 2193, South Africa

Penguin Books Ltd, Registered Offices: 80 Strand, London WC2R 0RL, England

www.penguin.com

First published in the United States of America by City Lights Books 2012
First published in Great Britain by Hamish Hamilton 2012
Published in Penguin Books 2012
001

The writings in this book are adapted from essays by Noam Chomsky
distributed by *The New York Times* Syndicate

Printed in England by Clays Ltd, St Ives plc

ISBN: 978-0-241-95258-0

www.greenpenguin.co.uk

MIX
Paper from
responsible sources
FSC™ C018179

Penguin Books is committed to a sustainable
future for our business, our readers and our planet.
This book is made from Forest Stewardship
Council™ certified paper.

ALWAYS LEARNING   **PEARSON**

We're an empire now, and when we act, we create our own reality. And while you're studying that reality—judiciously, as you will—we'll act again, creating other new realities, which you can study too, and that's how things will sort out. We're history's actors . . . and you, all of you, will be left to just study what we do.

—Senior adviser to former President George W. Bush, as quoted in the *New York Times Magazine*, October 17, 2004

Contents

Foreword: Remaking the Future
    by John Stickney, Senior Editor, New York
    Times Syndicate *11*

Threats, Talks and a Hoped-for Accord with North
    Korea *17*

Tortilla Wars *21*

We Own the World *25*

Gaza and the Future of a Palestinian-Israeli Peace *31*

Containing Iran *37*

Hypocrisies and Hopes in Annapolis *41*

The Somalia Syndrome *45*

"Good News" from Iraq, Afghanistan and Pakistan *51*

In the Campaign, the Unspeakable War *57*

Would a Democrat Change U.S. Middle East Policy? *63*

Delaying Doomsday: This Century's Challenges *73*

Middle East Road Trip *83*

Iraq Oil: A Deal With the Devil *87*

Nuclear Threats: All Options Are on the Table *93*

Georgia and the Neo-con Cold Warriors *99*

The Campaign and the Financial Crisis *105*

Challenges for Barack Obama: Part 1
    The Election and the Economy *111*

Challenges for Barack Obama: Part 2
    Iraq, Pakistan and Afghanistan *117*

Nightmare in Gaza *125*

Barack Obama and Israel-Palestine *131*

Latin America, Defiant *135*

Down with the Durand Line! *141*

A Tradition of Torture *145*

Obama on Israel-Palestine *151*

A Season of Travesties *155*

Making War to Bring Peace *161*

Militarizing Latin America *167*

War, Peace and Obama's Nobel *171*

The Legacy of 1989 in Two Hemispheres *177*

Presidential "Peacekeeping" in Latin America *183*

The Corporate Takeover of U.S. Democracy *189*

The Unelected "Architects of Policy" *195*

A "Regrettable" Event in East Jerusalem *201*

Rust Belt Rage *207*

The Real Threat Aboard the Freedom Flotilla *211*

Storm Clouds Over Iran *215*

The War in Afghanistan: Echoes of Vietnam *221*

China and the New World Order: Part 1 *225*

China and the New World Order: Part 2 *231*

The U.S. Elections: Outrage, Misguided *235*

The Charade of Israeli-Palestinian Talks *241*

Breaking the Israel-Palestine Deadlock *247*

The Arab Word Is on Fire *253*

The Cairo-Madison Connection *259*

Libya and the World of Oil *256*

The International Assault on Labor *269*

The Revenge Killing of Osama Bin Laden *275*

In Israel, a Tsunami Warning *281*

America in Decline *285*

After 9/11, Was War the Only Option? *291*

The Threat of Warships on an "Island of World
    Peace" *297*

Occupy the Future *301*

Index *307*

Foreword: Remaking the Future

By John Stickney, Senior Editor
The New York Times Syndicate

Who or what makes the future?

This book's title refers to the epigraph, which states the worldview of a former senior adviser to President George W. Bush:

"We're an empire now, and when we act, we create our own reality. And while you're studying that reality—judiciously, as you will—we'll act again, creating other new realities, which you can study too, and that's how things will sort out. We're history's actors . . . and you, all of you, will be left to just study what we do."

"History's actors" should be so unlucky that an observer as knowledgeable, indefatigable, writerly and unflinching as Noam Chomsky is on hand to "study what we do."

Events belie the senior adviser as this book goes to press. The Occupy movement is on fire worldwide, ignited by outrages that Chomsky explores here: inequality, disenfranchisement, official arrogance and deceit.

"I've never seen anything quite like the Occupy movement in scale and character, here and worldwide," Chomsky said at Occupy Boston on October 22, 2011, a talk adapted for this book's final entry. "The Occupy outposts are trying to create cooperative communities that just might be the basis for the kinds of lasting organizations necessary to overcome the barriers ahead and the backlash that's already coming."

Such is the material for the monthly column that Chomsky writes for the New York Times Syndicate. *Making the Future* comprises the second edition of those columns.

The first edition, *Interventions*, published in 2007, was banned at Guantánamo in 2009.

As Carol Rosenberg tells it in the *Miami Herald*, a Pentagon defense lawyer sent an Arabic-language edition of *Interventions* to a detainee. U.S military censors rejected it.

"This happens sometimes in totalitarian regimes," Chomsky told Rosenberg by email. "Of some incidental interest, perhaps, is the nature of the book they banned. It consists of op-eds written for the New York Times Syndicate and distributed by them. The subversive rot must run very deep."

Subverted or otherwise, I edit the Chomsky column. It began when Syndicate editors were looking for commentary on the first anniversary of the September 11 attacks. Chomsky's book *9-11*, published in October 2001, had become a best seller. He was a natural to approach.

Chomsky's first Syndicate op-ed, "9-11: Lessons Unlearned," gained a wide readership, especially abroad. Chomsky's writing was largely disregarded by the mainstream press at home, perhaps because of his uncompromising perspective on what he has called "the global hegemon." But his voice traveled across borders. Why not try a monthly column?

Taken together, the columns in this edition present a narrative of the events that have made the future since 2007: the wars in Afghanistan and Iraq; the U.S. presidential race; the ascendancy of China; Latin America's leftward turn; the threat of nuclear proliferation in Iran and North Korea; Israel's invasion of Gaza and expansion of settlements in Jerusalem and the West Bank; developments in climate change; the world financial crisis; the Arab Spring; the death of Osama bin Laden; and the Occupy protests.

As often happens, Chomsky's columns anticipate events. His August 2011 column, "America in Decline," foreshadows a premise of the Occupy movement:

"The resulting concentration of wealth [since the 1970s] yielded greater political power, accelerating a vicious cycle that has led to extraordinary wealth for a fraction of 1 percent of the population, mainly, while for the large majority real incomes have virtually stagnated."

In every sense Chomsky lives up to the title of "public intellectual." He is constantly on the road, giving talks. (Often on campuses he will speak to an audience on current events and to a smaller gathering on his day-job specialty, linguistics.) A hallmark of a Chomsky talk is the question-and-answer period afterward—which tends to continue, freewheeling, until organizers shepherd him to the next stop on the schedule.

Chomsky frequently grants interviews to periodicals around the world, and he maintains an exhaustive email correspondence. Meanwhile he reads voraciously: the mainstream press, journals, books and blogs from the United States and around the world. His office at Massachusetts Institute of Technology is in a building designed by Frank Gehry, where walls tilt inward as the horizontal stack of books on Chomsky's desk climbs ever upward.

Chomsky's habit with newspapers is to read down the

articles until he comes to the most revealing material, of-ten submerged toward the end—such as the quote from the unnamed Bush adviser.

This collection's op-eds very much reflect Chomsky's public exchanges, as at Occupy Boston. Notes for talks and answers to questions, along with his reading, evolve into columns and material for his books, and vice versa. The dialogue with all his audiences informs Chomsky's voice.

The op-ed "Making War to Bring 'Peace,'" from July 2009, tracked his speech that month to the U.N. General Assembly. As a panelist—uncharacteristically wearing a necktie—Chomsky spoke on the policy known as "respon-sibility to protect," or R2P. Afterward delegates lined up with questions and challenges.

Chomsky's outspokenness puts him at risk. In 2010, on his way from Jordan to the West Bank to give a talk at Birzeit University, Israeli officials barred Chomsky at the border. He gave the Birzeit talk anyway, in a videoconference.

Everywhere, Chomsky is besieged by his editors, myself included. Road-weary or jet-lagged, he heeds yet another deadline and delivers an op-ed, often after mid-night—carefully researched and annotated, as befits his scholarly background. Then he submits to the editing pro-cess, another back-and-forth in which his editors come to respect the care and craft and accuracy of his work.

In his Occupy Boston talk, Chomsky tells truths and summons his listeners to informed action:

"Karl Marx said, 'The task is not just to understand the world but to change it.' A variant to keep in mind is that if you want to change the world you'd better try to under-stand it. That doesn't mean listening to a talk or reading a book, though that's helpful sometimes. You learn from participating. You learn from others. You learn from the people you're trying to organize. We all have to gain the understanding and the experience to formulate and implement ideas."

How Chomsky thinks, and what he covers in this collection, engage readers not to leave to anybody but themselves the task of making the future.

Threats, Talks and a Hoped-for Accord with North Korea

APRIL 2, 2007

A truism in human as well as world affairs is that if you threaten people, they will defend themselves. If you reach out in good faith, people are likely to reach back.

A case in point is the long tortuous relationship between the United States and North Korea. One of many illustrations was when, in 2002, President Bush named North Korea a charter member of the "Axis of Evil." North Korea was developing plutonium bombs and represented an imminent threat, according to U.S. intelligence. The charges in fact instigated the very threats that Washington had warned against.

North Korea, unlike Iraq, could already defend itself—with massed artillery aimed at Seoul, South Korea, and at U.S. troops near the demilitarized zone. The stakes rose harrowingly as North Korea began amassing its nuclear weapons arsenal.

Then, in February of this year [2007], multilateral talks convened in Beijing (which included China, Japan, Russia and South Korea as well as North Korea and the United States). Within days, in an apparent about-face for both Pyongyang and Washington, the talks produced heartening results: North Korea, responding to conciliatory offers, agreed to start dismantling its nuclear facilities and allow nuclear inspectors back in the country.

The Bush administration declared the talks a success. The spin was that North Korea, faced with a potentially regime-changing isolation from the world community, had backed down. What actually happened is quite different, and instructive about how to help defuse the North Korea crisis and others like it.

Last October [2006], North Korea conducted a nuclear test in the mountains near the Chinese border, apparently a dud, yet with enough firepower to inch the world a bit farther toward nuclear Armageddon. Last July [2006], North Korea resumed long-range missile testing—also a fizzle, yet with the ominous signal that payload and delivery might eventually come together.

The test, and the missile firing, can be added to the record of the Bush administration's achievements.

Leon V. Sigal, one of the foremost experts on nuclear diplomacy in the region, sets the context. "When President Bush took office," Sigal wrote in the November 2006 issue of *Current History*, "the North had stopped testing longer-range missiles. It had one or two bombs' worth of plutonium and was verifiably not making more. Six years later, it has eight to 10 bombs' worth, has resumed longer-range missile tests, and feels little restraint about nuclear testing."

Reviewing the record, Sigal concludes that "Pyongyang in fact has been playing tit for tat—reciprocating whenever Washington cooperates and retaliating whenever Washington reneges—in an effort to end enmity."

An example for cooperation was set more than a decade ago. Haltingly and unevenly, the Clinton administration began a process to normalize U.S. political and economic relations with North Korea and guarantee its security as a non-nuclear state. In 1994, North Korea agreed not to enrich uranium.

Then, in 2002, Bush's "Axis of Evil" militarism had the predictable effects: North Korea returned to the development of missiles and nuclear weapons, expelled U.N. inspectors and withdrew from the Non-Proliferation Treaty.

Eventually, however, under pressure from Asian countries, the Bush administration agreed to talks, leading to an agreement in September 2005 that North Korea would abandon "all nuclear weapons and existing weapons programs" and allow inspections, in exchange for international aid with a light-water reactor, and a nonaggression pledge from the United States. By the agreement, the two sides would "respect each other's sovereignty, exist peacefully together and take steps to normalize relations."

Had that agreement been implemented, there would not have been a North Korean bomb test or the heightened conflict, always verging on the edge of nuclear war.

Much as in the case of Iran during the same years, the Bush administration chose confrontation over diplomacy, immediately undermining the accord. It disbanded the international consortium set up to provide the light-water reactor, renewed threats of force and pressured banks to freeze North Korea's hard-currency accounts. The grounds were that North Korea was using banks for illegal transfers, perhaps for counterfeiting, though the evidence is sketchy at best.

By February of this year [2007], "increasing pressure on the North Korean regime and an embattled U.S. administration seeking success in its dealings with one of the 'axes of evil' have helped breathe life into a process long considered terminal," Anna Fifield wrote in the *Financial Times*.

The new agreement is similar to the one that Washington had scuttled in 2005. Immediately after the new agreement was reached, Washington conceded that its 2002 charges against North Korea were based on dubious evidence. The Bush administration, notorious for fitting the facts to the policy in Iraq, may have also skewed the intelligence on North Korea.

"It is unclear why the new assessment is being disclosed now," David E. Sanger and William J. Broad wrote in the *New York Times*. "But some officials suggested that the timing could be linked to North Korea's recent agreement to reopen its doors to international arms inspectors. As a result, these officials have said, the intelligence agencies are facing the possibility that their assessments will once again be compared to what is actually found on the ground."

The lesson from the cycle of reciprocation and retaliation, talks and threats, is as Sigal observes: Diplomacy can work, if conducted in good faith.

Tortilla Wars

MAY 9, 2007

The chaos that derives from the so-called international order can be painful if you are on the receiving end of the power that determines that order's structure.

Even tortillas come into play in the evolving scheme of things. Recently, in many regions of Mexico, tortilla prices jumped by more than 50 percent. In January [2007], in Mexico City, tens of thousands of workers and farmers rallied in the Zócalo, the city's central square, to protest the skyrocketing cost of tortillas.

In response, the government of President Felipe Calderón cut a deal with Mexican producers and retailers to limit the price of tortillas and corn flour, very likely a temporary expedient.

In part the price-hike threat to the food staple for Mexican workers and the poor is a consequence of the U.S. stampede to corn-based ethanol as an energy substitute for oil, whose major wellsprings, of course, are in regions that are highly conflicted.

In the United States, too, the ethanol initiatives have raised food prices over a broad range, including other crops, livestock and poultry.

The connection between instability in the Middle East and the cost of feeding a family in the Americas isn't direct, of course. But as with all international trade, power tilts the balance. A leading goal of U.S. foreign policy has long been to create a global order in which U.S. corporations have free access to markets, resources and investment opportunities. The objective is commonly called "free trade," a posture that collapses quickly on examination.

It's not unlike what Britain, a predecessor in world domination, imagined during the latter part of the nineteenth century, when it embraced limited free trade, after 150 years of state intervention and violence had helped the nation achieve far greater industrial power than any rival.

The United States has followed much the same pattern. Generally, great powers are willing to enter into some limited degree of free trade when they're convinced that the economic interests under their protection are going to do well. That has been, and remains, a primary feature of the international order.

The ethanol boom fits the pattern. As discussed by agricultural economists C. Ford Runge and Benjamin Senauer in the current issue of *Foreign Affairs*, "the biofuel industry has long been dominated not by market forces but by politics and the interests of a few large companies," in large part Archer Daniels Midland, the major ethanol producer. Ethanol production is feasible thanks to substantial state subsidies and very high tariffs to exclude much cheaper and more efficient sugar-based Brazilian ethanol.

In March [2007], during President Bush's trip to Latin America, the one heralded achievement was a deal with Brazil on joint production of ethanol. But Bush, while spouting free-trade rhetoric for others in the conventional manner, emphasized forcefully that the high tariff to protect U.S. producers would remain, along with the many forms of government subsidy for the industry.

Despite the huge, taxpayer-supported agricultural subsidies, the prices of corn—and tortillas—have been climbing rapidly. One factor is that industrial users of imported U.S. corn increasingly purchase cheaper Mexican varieties used for tortillas, raising prices.

The 1994 U.S.-sponsored North American Free Trade Agreement (NAFTA) may also play a significant role, one that is likely to increase. An unlevel-playing-field impact of NAFTA was to flood Mexico with highly subsidized agribusiness exports, driving Mexican producers off the land.

Mexican economist Carlos Salas reviews data showing that after a steady rise until 1993, agricultural employment began to decline when NAFTA came into force, primarily among corn producers—a direct consequence of NAFTA, he and other economists conclude. One-sixth of the Mexican agricultural workforce has been displaced in the NAFTA years, a process that is continuing, depressing wages in other sectors of the economy and impelling emigration to the United States. Max Correa, secretary general of the group Central Campesina Cardenista, estimates that "for every five tons bought from foreign producers, one campesino becomes a candidate for migration."

It is, presumably, more than coincidental that President Clinton militarized the Mexican border, previously quite open, in 1994, along with implementation of NAFTA.

The "free trade" regime drives Mexico from self-sufficiency in food toward dependency on U.S. exports. And as the price of corn goes up in the United States, stimulated by corporate power and state intervention, one can anticipate that the price of staples may continue its sharp rise in Mexico.

Increasingly, biofuels are likely to "starve the poor" around the world, according to Runge and Senauer, as

staples are converted to ethanol production for the privileged—cassava in sub-Saharan Africa, to take one ominous
example. Meanwhile, in Southeast Asia, tropical forests are
cleared and burned for oil palms destined for biofuel, and
there are threatening environmental effects from input-
rich production of corn-based ethanol in the United States
as well.

The high price of tortillas and other, crueler vagaries
of the international order illustrate the interconnectedness
of events, from the Middle East to the Middle West, and
the urgency of establishing trade based on true democratic
agreements among people, and not interests whose principal hunger is for profit for corporate interests protected
and subsidized by the state they largely dominate, whatever the human cost.

We Own the World

JUNE 6, 2007

In crude and brutal societies, the Party Line is publicly proclaimed, and it must be obeyed, or else. What you believe is your own business, and of lesser concern.

In societies where the state has lost the capacity to control by force, the Party Line is not proclaimed. Rather, it is presupposed, and then vigorous debate is encouraged within the limits imposed by unstated doctrinal orthodoxy.

The crude system leads to natural disbelief. The sophisticated variant gives the impression of openness and freedom, and serves to instill the Party Line as beyond question, even beyond thought, like the air we breathe.

In the ever more precarious standoff between Washington and Tehran, one Party Line confronts another. Among the well-known immediate victims are the Iranian-American detainees Parnaz Azima, Haleh Esfandiari, Ali Shakeri and Kian Tajbakhsh. But the whole world is held hostage to the U.S.-Iran conflict, where, after all, the stakes are nuclear.

Unsurprisingly, President Bush's announcement of a "surge" in Iraq—in reaction to the call of most Americans for steps toward withdrawal, and the even stronger demands of the (irrelevant) Iraqis—was accompanied by ominous leaks about Iranian-based fighters and Iranian-made improvised explosive devices in Iraq aimed at disrupting Washington's mission to gain victory, which is (by definition) noble.

Then followed the predictable debate: The hawks say we have to take violent measures against such outside interference in Iraq. The doves counter that we must make sure the evidence is compelling. The entire debate can proceed without absurdity only on the tacit assumption that we own the world. Therefore interference is limited to those who impede our objectives in a country that we invaded and occupy.

What are the plans of the increasingly desperate clique that narrowly holds political power in the United States? Reports of threatening, off-the-record statements by staffers for Vice President Cheney have heightened fears of an expanded war.

"You do not want to give additional argument to new crazies who say, 'Let's go and bomb Iran,'" Mohamed El-Baradei, director-general of the International Atomic Energy Agency, told the BBC last month [May 2007]. "I wake up every morning and see 100 Iraqis, innocent civilians, are dying."

U.S. Secretary of State Condoleeza Rice, as against the "new crazies," is supposedly pursuing the diplomatic track with Tehran. But the Party Line holds, unchanged. In April [2007], Rice spoke about what she would say if she encountered her Iranian counterpart, Manouchehr Mottaki, at the international conference on Iraq at Sharm el Sheikh. "What do we need to do? It's quite obvious," Rice said. "Stop the flow of arms to foreign fighters; stop the flow of foreign fighters across the borders." She is referring, of course, to Iranian fighters and arms. U.S. fighters and arms are not "foreign" in Iraq. Or anywhere.

The tacit premise underlying her comment, and virtually all public discussion about Iraq (and beyond) is that we own the world. Do we not have the right to invade and destroy a foreign country? Of course we do. That's a given. The only question is: Will the surge work? Or some other tactic? Perhaps this catastrophe is costing us too much. And those are the limits of the debates among the presidential candidates, the Congress and the media, with rare exceptions. The basic issues are not discussable.

Doubtless Tehran merits harsh condemnation, certainly for severe domestic repression and the inflammatory rhetoric of President Mahmoud Ahmadinejad (who has little to do with foreign affairs). It is, however, useful to ask how Washington would act if Iran had invaded and occupied Canada and Mexico, overthrown the governments there, slaughtered scores of thousands of people, deployed major naval forces in the Caribbean and issued credible threats to destroy the United States if it did not immediately terminate its nuclear energy programs (and weapons). Would we watch quietly?

After the United States invaded Iraq, "had the Iranians not tried to build nuclear weapons, they would be crazy," said Israeli military historian Martin van Creveld.

Surely no sane person wants Iran (or anyone) to develop nuclear weapons. A reasonable solution to the crisis would permit Iran to develop nuclear energy, in accord with its rights under the Non-Proliferation Treaty, but not nuclear weapons. Is that outcome feasible? It would be, under one condition: that the United States and Iran were functioning democratic societies, in which public opinion has a significant impact on public policy, overcoming the huge gulf that now exists on many critical issues, including this one.

That reasonable solution has overwhelming support among Iranians and Americans, who agree quite generally on nuclear issues, according to recent polls by the Program on International Policy Attitudes at the University of Maryland. The Iranian-American consensus extends to complete elimination of nuclear weapons everywhere (82 percent of Americans), and if that cannot be achieved, a "nuclear-weapons-free zone in the Middle East that would include Islamic countries and Israel (71 percent of Americans)." To 75 percent of Americans, it is better to build relations with Iran rather than use threats of force.

These facts suggest a possible way to prevent the current crisis from exploding, perhaps even to World War III, as predicted by British military historian Correlli Barnett. That awesome threat might be averted by pursuing a familiar proposal: democracy promotion—at home, where it is badly needed.

Although we cannot carry out the project directly in Iran, we can act to improve the prospects for the courageous reformers and oppositionists who are seeking to achieve just that. They include people like Saeed Hajjarian, Nobel laureate Shirin Ebadi and Akbar Ganji, and those who as usual remain nameless, among them labor activists.

We can improve the prospects for democracy promotion in Iran by sharply reversing state policy here so that it reflects popular opinion. That would entail withdrawing the threats that are a gift to the Iranian hard-liners and are bitterly condemned for that reason by Iranians truly concerned with democracy promotion. We can act to open some space for those who are seeking to overthrow the reactionary and repressive theocracy from within, instead of undermining their efforts by threats and aggressive militarism.

Democracy promotion, while no panacea, would be a useful step toward helping the United States become a "responsible stakeholder" in the international order (to adopt the term used for adversaries), instead of being an object of fear and dislike throughout much of the world. Apart from being a value in itself, a functioning democracy at home holds promise for a simple recognition that we don't own the world, we share it.

Gaza and the Future of a Palestinian-Israeli Peace

JULY 16, 2007

The death of a nation is a rare and somber event. But the vision of a unified, independent Palestine threatens to be another casualty of a Hamas-Fatah civil war, stoked by Israel and its enabling ally the United States.

Last month's [June 2007] chaos may mark the beginning of the end of the Palestinian Authority. That might not be an altogether unfortunate development for Palestinians, given U.S.-Israeli programs of rendering it little more than a quisling regime to oversee these allies' utter rejection of a viable independent state.

The events in Gaza took place in a developing context. In January 2006, Palestinians voted in a carefully monitored election, pronounced to be free and fair by international observers, despite U.S.-Israeli efforts to swing the election toward their favorite, Palestinian Authority president Mahmoud Abbas and his Fatah party. But Hamas won a surprising victory.

The punishment of Palestinians for the crime of voting the wrong way was severe. With U.S. backing, Israel stepped up its violence in Gaza, withheld funds it was legally obligated to transmit to the Palestinian Authority, tightened its siege and even cut off the flow of water to the arid Gaza Strip.

The United States and Israel made sure that Hamas

would not have a chance to govern. They rejected Hamas's call for a long-term cease-fire to allow for negotiations on a two-state settlement along the lines of an international consensus that Israel and United States have opposed, in virtual isolation, for more than thirty years, with rare and temporary departures.

Meanwhile Israel stepped up its programs of annexation, dismemberment and imprisonment of the shrinking Palestinian cantons in the West Bank, always with U.S. backing despite occasional minor complaints, accompanied by the wink of an eye and munificent funding.

There is a standard operating procedure for overthrowing an unwanted government: Arm the military to prepare for a coup. Accordingly, Israel and its U.S. ally armed and trained Fatah to win by force what it lost at the ballot box, with a military coup in Gaza.

A detailed and documented account by David Rose in *Vanity Fair* was confirmed by Norman Olsen, who served for twenty-six years in the Foreign Service, including four years working in the Gaza Strip and four years at the U.S. Embassy in Tel Aviv, and then moved on to become associate coordinator for counterterrorism at the Department of State. Olsen and his son review the State Department efforts to ensure that their candidate, Abbas, would win in the January 2006 elections and, when these efforts failed, to incite a coup by Fatah strongman Muhammad Dahlan. But "Dahlan's thugs moved too soon," the Olsens write, and a Hamas preemptive strike undermined the coup attempt.

The United States also encouraged Abbas to amass power in his own hands, appropriate behavior in the eyes of Bush administration advocates of presidential dictatorship. The strategy backfired, but Israel and the United

States quickly moved to turn the failed coup to their benefit. They now have a pretext for tightening the stranglehold on the people of Gaza.

"To persist with such an approach under present circumstances is indeed genocidal, and risks destroying an entire Palestinian community that is an integral part of an ethnic whole," writes international law scholar Richard Falk, U.N. Special Rapporteur for Israel-Palestine.

The approach is to be pursued unless Hamas meets the three conditions imposed by the "international community"—a technical term referring to the U.S. government and whoever goes along with it. For Palestinians to be permitted to peek out of the walls of their Gaza dungeon, Hamas must recognize Israel, renounce violence and accept past agreements, in particular, the Road Map of the Quartet (the United States, Russia, the European Union and the United Nations).

The hypocrisy is stunning. Obviously, the United States and Israel do not recognize Palestine or renounce violence. Nor do they accept past agreements. While Israel formally accepted the Road Map, it attached fourteen reservations that eviscerate it. To take just the first, Israel demanded that for the process to commence and continue, the Palestinians must ensure full quiet, education for peace, cessation of incitement, dismantling of Hamas and other organizations, and other conditions; and even if they were to satisfy this virtually impossible demand, the Israeli cabinet proclaimed that "the road map will not state that Israel must cease violence and incitement against the Palestinians."

Israel's rejection of the Road Map, with U.S. support, is unacceptable to the Western self-image, so it has been suppressed. The facts finally broke into the mainstream with Jimmy Carter's book *Palestine: Peace not Apartheid*, which elicited a torrent of abuse and desperate and disgraceful efforts to discredit it, but no discussion of such revelations as these, as far as I could discover.

While now in a position to crush Gaza, Israel can also proceed, with U.S. backing, to implement its plans in the West Bank, expecting to have the tacit cooperation of Fatah leaders who will be rewarded for their capitulation. Among other steps, Israel began to release the funds—estimated at $600 million—that it had illegally frozen in reaction to the January 2006 election.

Ex–prime minister Tony Blair is now to ride to the rescue. To Lebanese political analyst Rami Khouri, "appointing Tony Blair as special envoy for Arab-Israeli peace is something like appointing the Emperor Nero to be the chief fireman of Rome." Blair is the Quartet's envoy only in name. The Bush administration made it clear at once that he is Washington's envoy, with a very limited mandate. Secretary of State Rice (and President Bush) retain unilateral control over the important issues, while Blair would be permitted to deal only with problems of institution-building.

As for the short-term future, the best case would be a two-state settlement, in accord with the international consensus. That is still by no means impossible. It is supported by virtually the entire world, including the majority of the U.S. population. It has come rather close, once, during the last month of Bill Clinton's presidency—the sole meaningful U.S. departure from extreme rejectionism during the past thirty years. In January 2001, the United States lent its support to the negotiations in Taba, Egypt, that nearly achieved such a settlement before they were called off by Israeli Prime Minister Ehud Barak.

In their final press conference, the Taba negotiators expressed hope that if they had been permitted to continue their joint work, a settlement could have been reached. The years since have seen many horrors, but the possibility remains. As for the likeliest scenario, it looks unpleasantly close to the worst case, but human affairs are not predictable: Too much depends on will and choice.

Containing Iran

AUGUST 20, 2007

In Washington a remarkable and ominous campaign is under way to "contain Iran," which turns out to mean "containing Iranian influence," in a confrontation that *Washington Post* correspondent Robin Wright calls "Cold War II."

The sequel bears close scrutiny as it unfolds under the direction of former Kremlinologists Condoleezza Rice and Robert M. Gates, according to Wright. Stalin had imposed an Iron Curtain to bar Western influence; Bush-Rice-Gates are imposing a Green Curtain to bar Iranian influence.

Washington's concerns are understandable. In Iraq, Iranian support is welcomed by much of the majority Shiite population. In Afghanistan, President Karzai describes Iran as "a helper and a solution." In Palestine, Iranian-backed Hamas won a free election, eliciting savage punishment of the Palestinian population by the United States and Israel for voting "the wrong way." In Lebanon, most Lebanese see Iranian-backed Hezbollah "as a legitimate force defending their country from Israel," Wright reports.

And the Bush administration, without irony, charges that Iran is "meddling" in Iraq, otherwise presumably free from foreign interference. The ensuing debate is partly technical. Do the serial numbers on the improvised explosive devices really trace back to Iran? If so, does the leadership of Iran know about the IEDs, or only the Iranian Revolutionary Guards? Settling the debate, the White House plans to brand the Revolutionary Guards as a "specially designated global terrorist" force, an unprecedented action against a national military branch, authorizing Washington to undertake a wide range of punitive actions.

The saber-rattling rhetoric about "containing Iran" has escalated to the point where both political parties and practically the whole U.S. press corps accept it as legitimate and, in fact, honorable. And also that "all options are on the table," to quote the leading presidential candidates—possibly even nuclear weapons.

The U.N. Charter outlaws "the threat or use of force." The United States, which has chosen to become an outlaw state, disregards international laws and norms. We're allowed to threaten anybody we want—and to attack anybody we want.

Cold War II also may bring about an arms race. The United States is proposing a $20 billion arms sale to Saudi Arabia and other Gulf states, while increasing annual military aid to Israel by 30 percent, to $30 billion over ten years. Egypt is down for a $14 billion, ten-year deal. The aim is to counter "what everyone in the region believes is a flexing of muscles by a more aggressive Iran," says an unnamed senior U.S. government official. Iran's "aggression" consists in its being welcomed within the region, and allegedly supporting resistance to U.S. forces in neighboring Iraq.

Unquestionably, Iran's government merits sharp criticism. The prospect that Iran might develop nuclear weapons is deeply troubling. Though Iran has every right to develop nuclear energy, no one—including the majority of

Iranians—wants it to have nuclear weapons. That would add to the much more serious dangers presented by its near neighbors Pakistan, India and Israel, all nuclear-armed with the blessing of the United States.

Iran resists U.S. or Israeli domination of the Middle East but scarcely poses a military threat. Any potential threat to Israel might be overcome if the United States would accept the view of the great majority of its own citizens and of Iranians and permit the Middle East to become a nuclear-weapons-free zone, including Iran and Israel, and U.S. forces deployed there. One may also remember that U.N. Security Council Resolution 687, of 1991, to which Washington and London appealed in their efforts to provide a thin legal cover for their invasion of Iraq, calls for "establishing in the Middle East a zone free from weapons of mass destruction and all missiles for their delivery."

Washington's feverish new Cold War "containment" policy has spread even to Europe. The United States wants to install a "missile defense system" in the Czech Republic and Poland that is being marketed to Europe as a shield against Iranian missiles. Even if Iran had nuclear weapons and long-range missiles, the chances of its using them to attack Europe are perhaps on a par with the chances of Europe's being hit by an asteroid. In any case, if Iran were to indicate the slightest intention of aiming a missile at Europe or Israel, the country would be vaporized.

Of course Russia is upset by the shield proposal. We can imagine how the United States would respond if a Russian anti-missile system were erected in Canada. The Russians have every reason to regard an anti-missile system as part of a first-strike weapon against them. As is well known, such a system could never impede a first strike, but it could conceivably impede a retaliatory strike. On all sides, "missile defense" is therefore understood to be a first-strike weapon, potentially eliminating a deterrent to attack. Even more obviously, the only military function of such a system with regard to Iran, the declared aim, would be to bar an Iranian deterrent to U.S. or Israeli aggression.

The shield, then, ratchets the threat of war a few notches higher, in the Middle East and elsewhere, with incalculable consequences, and the potential for a terminal nuclear war. The immediate fear is that by accident or design, Washington's war planners or their Israeli surrogate might decide to escalate their Cold War II into a hot one.

There are many nonmilitary measures to "contain" Iran, including a de-escalation of rhetoric and hysteria all around, and agreeing to negotiations in earnest for the first time—if indeed all options are on the table.

Hypocrisies and Hopes in Annapolis

NOVEMBER 8, 2007

The crimes against Palestinians in the Occupied Territories and elsewhere, particularly since the Palestinians voted "the wrong way" in the Hamas victory last year [2006], are so shocking that the only emotionally valid reaction is rage and a call for extreme actions. But that does not help the victims, and is likely to harm them. Our actions have to be adapted to real-world circumstances, difficult as it may be to stay calm in the face of shameful crimes, in which we in the United States are directly and crucially implicated.

We are approaching President Bush's Annapolis conference on Israel-Palestine, the administration's first potentially serious diplomatic initiative in that conflict.

Ideally, the Annapolis negotiations should begin at the point that had been reached in Taba, Egypt, in January 2001. That week was the one moment in thirty years when the United States and Israel abandoned the rejectionist stance that they have maintained in virtual isolation until the present. And Taba may have come close to a possible two-state settlement, with a reasonable land-swap. The conventional fabrication is that at Taba the Palestinians rejected Israel's generous offer. In fact, the conference was terminated abruptly by Israeli Prime Minister Ehud Barak, at a moment when negotiators reported that they were close to agreement.

Perhaps Taba nearly succeeded because the United

States was not there as a mediator. Washington's policy toward Israel-Palestine has long been at odds with conventional images. "Every [U.S.] administration since 1967, when Israel won a war and occupied the West Bank and the Gaza Strip, has privately favored returning almost all of that territory to the Palestinians for the purposes of creating a separate Palestinian state," Leslie Gelb, the respected policy analyst, observed two months ago [September 2007] in the *New York Times Book Review*. Note the word "privately." Why not publicly?

Gelb cannot have meant that the difference in stance came from fear of the terrifying Israel lobby, since he is at pains to deny the thesis that it is a powerful and intimidating force. So why only "privately"? Perhaps because such an interpretation supports the comforting self-image of the United States as an "honest broker," thwarted in our noble efforts by violent and irrational foreigners, with Palestinians assigned the leading role in the drama.

We know what administrations have said publicly. They have rejected anything remotely of the sort, ever since 1976, when the United States vetoed a Security Council resolution calling for a two-state settlement on the international border (the Green Line), incorporating all the relevant wording of U.N. Resolution 242, of November 1967.

By now virtually the entire world agrees on the two-state international consensus, along the lines almost agreed upon at Taba. That includes all the Arab states, who go on to call for full normalization of relations with Israel. It includes Iran, which accepts the Arab League position. It includes Hamas, whose leaders have repeatedly and publicly called for a two-state settlement, even in the U.S. press.

It includes even Hamas's most militant figure, Khaled Meshal, in exile in Syria.

Israel has consistently rejected the international consensus, and the United States backs that rejection fully. Bush II has gone to new extremes in rejectionism, declaring that the illegal West Bank settlements must remain part of Israel. But the Party Line remains undisturbed: Bush, Condoleezza Rice and the rest are yearning to realize Bush's "vision" of a Palestinian state, persisting in the noble endeavor of the longtime "honest broker."

Rejectionism goes far beyond words. More significant are actions on the ground: settlement programs, the annexation wall, closures, checkpoints and much worse. The story continues as the Annapolis conference approaches. Just one example: Israel has just confiscated more Arab land to build a bypass road for Palestinians in order to "push the Palestinian traffic between Bethlehem and Ramallah deep into the desert and effectively bar [Palestinians] from the central part of the West Bank," part of the E-1 development project, east of Jerusalem, designed to incorporate the town of Ma'aleh Adumim within Israel and, in effect, to bisect the West Bank, according to the Israeli peace organization Gush Shalom. "With such policies enacted by the government, the famous Annapolis Conference is emptied of all meaning, long before it convenes."

No realistic proposal has been advanced that does not take a two-state settlement at least as a starting point, along the general lines of Taba. Informal negotiations followed, leading to several detailed proposals, notably the Geneva Accord of December 2003, applauded by most of the world but dismissed by the "boss-man called partner," as Israeli political analyst Amir Oren describes the U.S.-

Israel relationship. Without U.S. support, Israel cannot achieve its expansionist aims, which lays the responsibility on us here in the United States.

Plenty of pitfalls are ahead. Some of Ehud Olmert's closest advisers have endorsed a version of the "land-swap" policy of the ultra-right Yisrael Beitenu leader Avigdor Lieberman. Such a swap would give the Palestinians technical authority over the heavily Arab "triangle" region in Israel, bordering the Green Line. In exchange, Israel would annex the West Bank settlements that encompass precious water resources and valuable land, leaving the rest cantonized and imprisoned, with the Israeli takeover of the Jordan Valley. The inhabitants, of course, are not to be consulted.

In the coming weeks, and the longer term, there is plenty of educational and organizing work to be done, among an American population that is largely receptive, though deluged with propaganda and deceit. It will not be easy. It never is. But much harder tasks have been accomplished with dedicated and persistent effort.

The Somalia Syndrome

DECEMBER 17, 2007

"This poor country keeps taking one blow after another," Peter Goossens observed two months ago [October 2007] in an interview with the *New York Times'* Jeffrey Gettleman. "Ultimately, it will break." The country is Somalia, and Goossens directs the World Food Program, which is now feeding some 1.2 million people there, 15 percent of the population.

This tragic and tortured land is "marching right up to the edge of a crisis," Goossens said. "Any additional little thing, any little flood or drought, will push them over."

Somalia, war- and famine-torn, is beset from within and without. With a vigilance stepped-up since September 11, 2001, the United States has reformulated its long-standing efforts to control the Horn of Africa (Djibouti, Ethiopia, Eritrea and Somalia) as a front line in the "war on terror," and Somalia is at its very tip. The current crisis in Somalia may be regarded partly as collateral damage from that "war on terror" and the geopolitical concerns reframed in these terms.

As Somalia sinks deeper into chaos, members of the African Union have sent small peacekeeping forces there, and pledged to send more if funding is made available. But they are unlikely to do so, "because there is no peace to keep [in Somalia] in the first place," Richard Cornwell, of the Institute for Security Studies in South Africa, told

Scott Baldauf and Alexis Okeowo of the *Christian Science Monitor* in May [2007].

By November [2007], the United Nations noted that Somalia had "higher malnutrition rates, more current bloodshed and fewer aid workers than Darfur," Gettleman reported. Ahmedou Ould-Abdallah, the top U.N. official for Somalia, described its plight as "the worst on the continent."

The United Nations, however, lacks the capacity to reach the people who are hungry, exposed, sick and dying in Somalia, according to Eric Laroche, head of U.N. humanitarian operations there.

"If this were happening in Darfur, there would be a big fuss," Laroche said. "But Somalia has been a forgotten emergency for years."

One distinction, hard to miss, is that the tragedy of Darfur can be blamed on someone else, in fact an official enemy—the government of Sudan and its Arab militias—while responsibility for the current disaster in Somalia, like others there that preceded it, lies substantially in our own hands.

In 1992, after the overthrow of the Somali dictatorship by clan-based militias and the ensuing famine, the United States sent thousands of soldiers on a dubious "rescue mission" to assist with humanitarian operations. In October 1993, during the "Battle of Mogadishu," two Black Hawk helicopters were shot down by Somali militiamen, leaving eighteen U.S. Army Rangers dead, along with perhaps a thousand Somalis.

U.S. forces were immediately withdrawn in a manner that continued the murderous ratio. "In the final stages of the troops' retreat, every bullet fired against them was an-

swered, it seemed, by 100," *Los Angeles Times* correspondent John Balzar reported. As for the Somali casualties,
Marine Lt. Gen. Anthony Zinni, who commanded the operation, informed the press that "I'm not counting bodies.
. . . I'm not interested."

CIA officials privately conceded that during the U.S.
operations in Somalia, in which thirty-four U.S. soldiers
were lost, Somali casualties—militiamen and civilians—
may have been 7,000 to 10,000, Charles William Maynes
reported in *Foreign Policy*.

The "rescue mission," which may have killed about as
many Somalis as it saved, left the country in the hands of
brutal warlords.

"After that, the United States—and much of the rest
of the world—basically turned its back on Somalia," Gettleman reports. "But in the summer of 2006, the world
started paying attention again after a grassroots Islamist
movement emerged from the clan chaos and seized control
of much of the country," leaving only an enclave adjoining
Ethiopia in the hands of the Western-recognized Transitional Federal Government.

During their brief tenure, the Islamists "didn't cause us
any problems," Laroche reports. Ould-Abdallah called the
six months of their rule Somalia's "golden era," the only
period of peace in Somalia for years. Other U.N. officials
concur, observing that "the country was in better shape
during the brief reign of Somalia's Islamist movement last
year" than it has been since Ethiopia invaded in December 2006 to impose the rule of the Transitional Federal
Government.

The Ethiopian invasion, with U.S. backing and direct participation, took place immediately after the U.N.

Security Council, at U.S. initiative, passed Resolution 1725 for Somalia, which called upon all states "to refrain from action that could provoke or perpetuate violence and violations of human rights, contribute to unnecessary tension and mistrust, endanger the ceasefire and political process, or further damage the humanitarian situation."

The invasion by Somalia's historical enemy, Christian Ethiopia, soon elicited a bitter resistance, leading to the present crisis.

The official reason for U.S. participation in Ethiopia's overthrow of the Islamist regime is the "war on terror"—which itself has engendered terror, quite apart from its own atrocities. Furthermore, the roots of the Islamic fundamentalist regime trace back to earlier stages of the "war on terror."

Immediately after September 11, 2001, the United States spearheaded an international effort to close down Al-Barakaat—a Dubai-based Somali remittance network that also runs major businesses in Somalia—on grounds that it was financing terror. This move was hailed by government and media as one of the great successes of the "war on terror." In contrast, Washington's withdrawal of its charges as without merit a year later aroused little interest.

The greatest impact of the closing of Al-Barakaat was in Somalia. According to the United Nations, in 2001 the enterprise was responsible for about half the $500 million in remittances to Somalia, "more than it earns from any other economic sector and 10 times the amount of foreign aid (Somalia) receives."

Al-Barakaat also played a major role in the economy, Ibrahim Warde observes in *The Price of Fear*, his devastating study of Bush's "financial war on terror." The frivolous attack on a very fragile society "may have played a role in the rise . . . of Islamic fundamentalists," Warde concludes—another familiar consequence of the "war on terror."

The renewed torture of Somalia falls within the context of U.S. efforts to gain firm control over the Horn of Africa, where the United States is launching a new Africa command and extending naval operations in crucial shipping lanes, part of the broader campaign to ensure its domination of the world's primary energy resources in the Gulf region and in Africa as well.

Just after World War II, when State Department planners were assigning each part of the world its "function" within the overall system of U.S. domination, Africa was considered unimportant. George Kennan, head of the State Department's Policy Planning Staff, advised that Africa should be handed over to Europe to "exploit" for its reconstruction. No longer. The resources of Africa are too valuable to be left to others, particularly with China extending its commercial reach.

If poor Somalia collapses in starvation and misery, that is merely a sideshow of grand geopolitical designs, and of little moment.

"Good News" from Iraq, Afghanistan and Pakistan

JANUARY 22, 2008

The U.S. occupying army in Iraq (euphemistically called the Multi-National Force–Iraq) carries out extensive studies of popular attitudes. Its December 2007 report of a study of focus groups was uncharacteristically upbeat.

The report concluded that the survey "provides very strong evidence" to refute the common view that "national reconciliation is neither anticipated nor possible." On the contrary, the survey found that a sense of "optimistic possibility permeated all focus groups . . . and far more commonalities than differences are found among these seemingly diverse groups of Iraqis."

This discovery of "shared beliefs" among Iraqis throughout the country is "good news, according to a military analysis of the results," Karen DeYoung reports in the *Washington Post*.

The "shared beliefs" were identified in the report. To quote DeYoung, "Iraqis of all sectarian and ethnic groups believe that the U.S. military invasion is the primary root of the violent differences among them, and see the departure of 'occupying forces' as the key to national reconciliation."

So, according to Iraqis, there is hope of national reconciliation if the invaders, responsible for the internal violence, withdraw and leave Iraq to Iraqis.

The report did not mention other good news: Iraqis

appear to accept the highest values of Americans, as established at the Nuremberg Tribunal—specifically, that aggression—"invasion by its armed forces" by one state "of the territory of another state"—is "the supreme international crime differing only from other war crimes in that it contains within itself the accumulated evil of the whole." The chief U.S. prosecutor at Nuremberg, Supreme Court Justice Robert Jackson, forcefully insisted that the Tribunal would be mere farce if we do not apply its principles to ourselves.

Unlike Iraqis, the United States, indeed the West generally, rejects the lofty values professed at Nuremberg, an interesting indication of the substance of the famous "clash of civilizations."

More good news was reported by General David Petraeus and Ambassador to Iraq Ryan Crocker during the extravaganza staged on September 11, 2007. Only a cynic might imagine that the timing was intended to insinuate the Bush-Cheney claims of links between Saddam Hussein and Osama bin Laden, so that by committing the "supreme international crime" they were defending the world against terror—which increased sevenfold as a result of the invasion, according to an analysis last year by terrorism specialists Peter Bergen and Paul Cruickshank.

Petraeus and Crocker provided figures to show that the Iraqi government was greatly accelerating spending on reconstruction, reaching a quarter of the funding set aside for that purpose. Good news indeed, until it was investigated by the Government Accountability Office, which found that the actual figure was one-sixth what Petraeus and Crocker reported, a 50 percent decline from the preceding year.

More good news is the decline in sectarian violence, attributable in part to the success of the murderous ethnic cleansing that Iraqis blame on the invasion; there are fewer targets for sectarian killing. But it is also attributable to Washington's decision to support the tribal groups that had organized to drive out Iraqi al-Qaida, and to an increase in U.S. troops.

It is possible that Petraeus's strategy may approach the success of the Russians in Chechnya, where fighting is now "limited and sporadic, and Grozny is in the midst of a building boom" after having been reduced to rubble by the Russian attack, C.J. Chivers reports in the *New York Times* last September [2007].

Perhaps someday Baghdad and Falluja too will enjoy "electricity restored in many neighborhoods, new businesses opening and the city's main streets repaved," as in booming Grozny. Possible, but dubious, considering the likely consequence of creating warlord armies that may be the seeds of even greater sectarian violence, adding to the "accumulated evil" of the aggression.

Iraqis are not alone in believing that national reconciliation is possible. A Canadian-run poll found that Afghans are hopeful about the future and favor the presence of Canadian and other foreign troops—the "good news," that made the headlines.

The small print suggests some qualifications. Only 20 percent "think the Taliban will prevail once foreign troops leave." Three-quarters support negotiations between the U.S.-backed Karzai government and the Taliban, and over half favor a coalition government. The great majority therefore strongly disagree with the U.S.-Canadian stance, and believe that peace is possible with a turn toward peaceful means. Though the question was not asked in the poll, it seems a reasonable surmise that the foreign presence is favored for aid and reconstruction.

There are, of course, numerous questions about polls in countries under foreign military occupation, particularly in places like southern Afghanistan. But the results of the Iraq and Afghan studies conform to earlier ones, and should not be dismissed.

Recent polls in Pakistan also provide "good news" for Washington. Fully 5 percent favor allowing U.S. or other foreign troops to enter Pakistan "to pursue or capture al-Qaida fighters." Nine percent favor allowing U.S. forces "to pursue and capture Taliban insurgents who have crossed over from Afghanistan."

Almost half favor allowing Pakistani troops to do so. And only a little more than 80 percent regard the U.S. military presence in Asia and Afghanistan as a threat to Pakistan, while an overwhelming majority believe that the United States is trying to harm the Islamic world.

The good news is that these results are a considerable improvement over October 2001, when a *Newsweek* poll found that "Eighty-three percent of Pakistanis surveyed say they side with the Taliban, with a mere 3 percent expressing support for the United States," and over 80 percent described Osama bin Laden as a guerrilla and 6 percent a terrorist.

Amid the outpouring of good news from across the region, there is now much earnest debate among political candidates, government officials and commentators concerning the options available to the United States in Iraq. One voice is consistently missing: that of Iraqis. Their "shared beliefs" are well known, as in the past. But they cannot be permitted to choose their own path any more than young children can. Only the conquerors have that right.

Perhaps here too there are some lessons about the "clash of civilizations."

In the Campaign,
the Unspeakable War

Iraq remains a significant concern for the population, but that is a matter of little moment in a modern democracy.

Not long ago, it was taken for granted that the Iraq war would be the central issue in the presidential campaign, as it was in the midterm election of 2006. But it has virtually disappeared, eliciting some puzzlement. There should be none.

The *Wall Street Journal* came close to the point in a front-page article on Super Tuesday, the day of many primaries: "Issues Recede in '08 Contest As Voters Focus on Character." To put it more accurately, issues recede as candidates, party managers and their public relations agencies focus on character. As usual. And for sound reasons. Apart from the irrelevance of the population, they can be dangerous.

Progressive democratic theory holds that the population—"ignorant and meddlesome outsiders"—should be "spectators," not "participants" in action, as Walter Lippmann wrote.

The participants in action are surely aware that on a host of major issues, both political parties are well to the right of the general population, and that public opinion is quite consistent over time, a matter reviewed in the useful study *The Foreign Policy Disconnect: What Americans Want from Our Leaders but Don't Get*, by Benjamin Page and Marshall Bouton. It is important, then, for the attention of the people to be diverted elsewhere.

The real work of the world is the domain of an enlightened leadership. The common understanding is revealed more in practice than in words, though some do articulate it: President Woodrow Wilson, for example, held that an elite of gentlemen with "elevated ideals" must be empowered to preserve "stability and righteousness," essentially the perspective of the Founding Fathers. In more recent years the gentlemen are transmuted into the "technocratic elite" and "action intellectuals" of Camelot, "Straussian" neocons of Bush II or other configurations.

For the vanguard who uphold the elevated ideals and are charged with managing the society and the world, the reasons for Iraq's drift off the radar screen should not be obscure. They were cogently explained by the distinguished historian Arthur M. Schlesinger Jr., articulating the position of the doves forty years ago when the U.S. invasion of South Vietnam was in its fourth year and Washington was preparing to add another 100,000 troops to the 175,000 already tearing South Vietnam to shreds.

By then the invasion launched by President Kennedy was facing difficulties and imposing difficult costs on the United States, so Schlesinger and other Kennedy liberals were reluctantly beginning to shift from hawks to doves.

In 1966, Schlesinger wrote that of course "we all pray" that the hawks are right in thinking that the surge of the day will be able to "suppress the resistance," and if it does, "we may all be saluting the wisdom and statesmanship of the American government" in winning victory while leaving "the tragic country gutted and devastated by bombs, burned by napalm, turned into a wasteland by chemical defoliation, a land of ruin and wreck," with its "political and institutional fabric" pulverized. But escalation probably won't succeed, and will prove to be too costly for ourselves, so perhaps strategy should be rethought.

As the costs to ourselves began to mount severely, it soon turned out that everyone had always been a strong opponent of the war (in deep silence).

Elite reasoning, and the accompanying attitudes, carry over with little change to commentary on the U.S. invasion of Iraq today. And although criticism of the Iraq war is far greater and more far-reaching than in the case of Vietnam at any comparable stage, nevertheless the principles that Schlesinger articulated remain in force in media and commentary.

It is of some interest that Schlesinger himself took a very different position on the Iraq invasion, virtually alone in his circles. When the bombs began to fall on Baghdad, he wrote that Bush's policies are "alarmingly similar to the policy that imperial Japan employed at Pearl Harbor, on a date which, as an earlier American president said it would, lives in infamy. Franklin D. Roosevelt was right, but today it is we Americans who live in infamy."

That Iraq is "a land of ruin and wreck" is not in question. Recently the British polling agency Oxford Research Business updated its estimate of extra deaths resulting from the war to 1.03 million—excluding Karbala and Anbar provinces, two of the worst regions. Whether that estimate is correct, or much overstated as some claim, there is no doubt that the toll is horrendous. Several million people are internally displaced. Thanks to the generosity of Jordan and Syria, the millions of refugees fleeing the wreckage of Iraq, including most of the professional classes, have not been simply wiped out.

But that welcome is fading, for one reason because Jordan and Syria receive no meaningful support from the perpetrators of the crimes in Washington and London; the idea that they might admit these victims, beyond a trickle, is too outlandish to consider.

Sectarian warfare has devastated Iraq. Baghdad and other areas have been subjected to brutal ethnic cleansing and left in the hands of warlords and militias, the primary thrust of the current counterinsurgency strategy developed by General Petraeus, who won his fame by pacifying Mosul, now the scene of some of the most extreme violence.

One of the most dedicated and informed journalists who have been immersed in the shocking tragedy, Nir Rosen, recently published an epitaph, "The Death of Iraq," in *Current History*.

"Iraq has been killed, never to rise again," Rosen writes. "The American occupation has been more disastrous than that of the Mongols, who sacked Baghdad in the 13th century"—a common perception of Iraqis as well. "Only fools talk of 'solutions' now. There is no solution. The only hope is that perhaps the damage can be contained."

Catastrophe notwithstanding, Iraq remains a marginal issue in the presidential campaign. That is natural, given the spectrum of hawk-dove elite opinion. The liberal doves adhere to their traditional reasoning and attitudes, praying that the hawks will be right and that the United States will win a victory in the land of ruin and wreck, establishing "stability," a code word for subordination to

Washington's will. By and large hawks are encouraged, and doves silenced, by the upbeat post-surge reports of reduced casualties.

As discussed earlier, in December [2007], the Pentagon released "good news" from Iraq, a study of focus groups from all over the country that found that Iraqis have "shared beliefs," so that reconciliation should be possible, contrary to claims of critics of the invasion. The shared beliefs were two. First, the U.S. invasion is the cause of the sectarian violence that has torn Iraq to shreds. Second, the invaders should withdraw and leave Iraq to its people.

A few weeks after the Pentagon report, *New York Times* military-Iraq expert Michael R. Gordon wrote a reasoned and comprehensive review of the options on Iraq policy facing the candidates for the presidential election. One voice is missing in the debate: Iraqis'. Their preference is not rejected. Rather, it is not worthy of mention. And it seems that there is no notice of the fact. That makes sense on the usual tacit assumption of almost all discourse on international affairs: We own the world, so what does it matter what others think? They are "unpeople," to borrow the term used by British diplomatic historian Mark Curtis in his work on Britain's crimes of empire.

Routinely, Americans join Iraqis in unpeoplehood. Their preferences too provide no options.

Would a Democrat Change U.S. Middle East Policy?[1]

MARCH 28, 2008

Recently, when Vice President Cheney was asked by ABC News correspondent Martha Raddatz about polls showing that an overwhelming majority of U.S. citizens oppose the war in Iraq, he replied, "So?"

"So—you don't care what the American people think?" Raddatz asked.

"No," Cheney replied, and explained, "I think you cannot be blown off course by the fluctuations in public opinion polls."

Later, White House spokeswoman Dana Perino, explaining Cheney's comments, was asked whether the public should have "input."

Her reply: "You had your input. The American people have input every four years, and that's the way our system is set up."

That's correct. Every four years the American people can choose between candidates whose views they reject, and then they should shut up.

Evidently failing to understand democratic theory, the public strongly disagrees.

"Eighty-one percent say when making 'an important decision' government leaders 'should pay attention to public opinion polls because this will help them get a sense of the public's views,'" reports the Program on International Policy Attitudes (PIPA), in Washington.

And when asked "whether they think that 'elections are the only time when the views of the people should have influence, or that also between elections leaders should consider the views of the people as they make decisions,' an extraordinary 94 percent say that government leaders should pay attention to the views of the public between elections."

The same polls reveal that the public has few illusions about how their wishes are heeded: 80 percent "say that this country is run by a few big interests looking out for themselves," not "for the benefit of all the people."

With its unbounded disregard for public opinion, the Bush administration has been well to the radical nationalist and adventurist extreme of the policy spectrum, and was subjected to unprecedented mainstream criticism for that reason.

A Democratic candidate is likely to shift more toward the centrist norm. However, the spectrum is narrow. Looking at the campaign records and statements of Hillary Clinton and Barack Obama, it is hard to see much reason to expect significant changes in policy in the Middle East.

IRAQ

It is important to bear in mind that neither Democratic candidate has expressed a principled objection to the invasion of Iraq. By that I mean the kind of objection that was universally expressed when the Russians invaded Afghanistan or when Saddam Hussein invaded Kuwait: condemnation on the grounds that aggression is a crime—in fact the "supreme international crime," as the Nuremberg

Tribunal determined. No one criticized those invasions merely as a "strategic blunder" or as involvement in "another country's civil war, a war [they] can't win" (statements that Obama and Clinton, respectively, later made about the Iraq invasion).

The criticism of the Iraq war is on grounds of cost and failure, what are called "pragmatic reasons," a stance that is considered hardheaded, serious, moderate—in the case of Western crimes.

The intentions of the Bush administration, and presumably McCain, were outlined in a Declaration of Principles released by the White House in November 2007, an agreement between Bush and the U.S.-backed Nuri al-Maliki government of Iraq.

The Declaration allows U.S. forces to remain indefinitely to "deter foreign aggression" (though the only threat of aggression in the region is posed by the United States and Israel, presumably not the intention) and for internal security, though not, of course, internal security for a government that would reject U.S. domination. The Declaration also commits Iraq to facilitate and encourage "the flow of foreign investments to Iraq, especially American investments"—an unusually brazen expression of imperial will, forcefully reiterated in another of Bush's signing statements the following January.

In brief, Iraq is to remain a client state, agreeing to allow permanent U.S. military installations (called "enduring" in the preferred Orwellism) and ensuring U.S. investors priority in accessing its huge oil resources—a reasonably clear statement of goals of the invasion that were evident all along to anyone not blinded by official doctrine.

What are the alternatives of the Democrats? They

were clarified in March 2007, when the House and Senate approved Democratic proposals setting deadlines for withdrawal. General Kevin Ryan (retired), senior fellow at Harvard University's Belfer Center of International Affairs, analyzed the proposals for the *Boston Globe*.

The proposals permit the president to waive their restrictions in the interests of "national security," which leaves the door wide open, Ryan writes. They permit troops to remain in Iraq "as long as they are performing one of three specific missions: protecting U.S. facilities, citizens or forces; combating al-Qaida or international terrorists; and training Iraqi security forces."

The facilities include the huge U.S. military bases being built around the country and the U.S. Embassy—actually a self-contained city within a city, unlike any embassy in the world. None of these major construction projects are under way with the expectation that they will be abandoned.

The other conditions are also open-ended. "The proposals are more correctly understood as a re-missioning of our troops," Ryan sums up: "Perhaps a good strategy—but not a withdrawal."

It is difficult to see much difference between the March 7 Democratic proposals and those of Obama and Clinton.

IRAN

With regard to Iran, Obama is considered more moderate than Clinton, and his leading slogan is "change." So let us keep to him.

Obama calls for more willingness to negotiate with Iran, but within the standard constraints. His reported position is that he "would offer economic inducements and a possible promise not to seek 'regime change' if Iran stopped meddling in Iraq and cooperated on terrorism and nuclear issues," and stopped "acting irresponsibly" by supporting Shiite militant groups in Iraq.

Some obvious questions come to mind. For example, how would we react if Iran's President Mahmoud Ahmadinejad said he would offer a possible promise not to seek "regime change" in Israel if it stopped its illegal activities in the occupied territories and cooperated on terrorism and nuclear issues?

Obama's moderate approach is well to the militant side of public opinion—a fact that passes unnoticed, as is often the case. Like all other viable candidates, Obama has insisted throughout the electoral campaign that the United States must threaten Iran with attack (the standard phrase is "keep all options open"), a violation of the U.N. Charter, if anyone cares. But a large majority of Americans have disagreed: 75 percent favor building better relations with Iran, as compared with 22 percent who favor "implied threats," according to PIPA.

All the surviving candidates, then, are opposed by three-fourths of the public on this issue.

American and Iranian opinion on the core issue of nuclear policy has been carefully studied. In both countries, a large majority holds that Iran should have the rights of any signer of the Nonproliferation Treaty: to develop nuclear power but not nuclear weapons.

The same large majorities favor establishing a "nuclear-weapons-free zone in the Middle East that would include both Islamic countries and Israel." More than 80 percent of Americans favor eliminating nuclear weapons altogether—a legal obligation of the states with nuclear weapons, officially rejected by the Bush administration.

And surely Iranians agree with Americans that Washington should end its military threats and turn toward normal relations.

At a forum in Washington when the PIPA polls were released in January 2007, Joseph Cirincione, senior vice president for National Security and International Policy at the Center for American Progress (and Obama adviser), said the polls showed "the common sense of both the American people and the Iranian people, (who) seem to be able to rise above the rhetoric of their own leaders to find common sense solutions to some of the most crucial questions" facing the two nations, favoring pragmatic diplomatic solutions to their differences.

Though we do not have internal records, there is good reason to believe that the Pentagon is opposed to an attack on Iran. The March 11 [2008] resignation of Admiral William Fallon as head of the Central Command, responsible for the Middle East, was widely interpreted to trace to his opposition to an attack, probably shared with the military command generally.

The December 2007 National Intelligence Estimate reporting that Iran had not pursued a nuclear weapons program since 2003, when it sought and failed to reach a comprehensive settlement with the United States, perhaps reflects opposition of the intelligence community to military action.

There are many uncertainties. But it is hard to see concrete signs that a Democratic presidency would improve the situation very much, let alone bring policy into line with American or world opinion.

ISRAEL-PALESTINE

On Israel-Palestine too, the candidates have provided no reason to expect any constructive change.

On his website, Obama, the candidate of "change" and "hope," states that he "strongly supports the U.S.-Israel relationship [and] believes that our first and incontrovertible commitment in the Middle East must be to the security of Israel, America's strongest ally in the Middle East."

Transparently, it is the Palestinians who face by far the most severe security problem, in fact a problem of survival. But Palestinians are not a "strong ally." At most, they might be a very weak one. Hence their plight merits little concern, in accord with the operative principle that human rights are largely determined by contributions to power, profit and ideological needs.

Obama's website presents him as a superhawk on Israel. "He believes that Israel's right to exist as a Jewish state should never be challenged." He is not on record as demanding that the right of countries to exist as Muslim (Christian, White) states "should never be challenged."

Obama calls for increasing foreign aid "to ensure that (the) funding priorities (for military and economic assistance to Israel) are met." He also insists forcefully that the United States must not "recognize Hamas unless it renounce[s] its fundamental mission to eliminate Israel." No state can recognize Hamas, a political party, so what he must be referring to is the government formed by Hamas after a free election that came out "the wrong way" and is therefore illegitimate, in accord with prevailing elite concepts of "democracy."

And it is considered irrelevant that Hamas has repeatedly called for a two-state settlement in accord with the international consensus, which the United States and Israel reject.

Obama does not ignore Palestinians: "Obama believes that a better life for Palestinian families is good for both Israelis and Palestinians." He also adds a reference to two states living side by side that is vague enough to be unproblematic to U.S. and Israeli hawks.

For Palestinians, there are now two options. One is that the United States and Israel will abandon their unilateral rejectionism of the past thirty years and accept the international consensus on a two-state settlement, in accord with international law and, incidentally, in accord with the wishes of a large majority of Americans. That is not impossible, though the two rejectionist states are working hard to render it so.

A second possibility is the one that the United States and Israel are actually implementing. Palestinians will be consigned to their Gaza prison and to West Bank cantons, virtually separated from one another by Israeli settlements and huge infrastructure projects, the whole imprisoned as Israel takes over the Jordan Valley.

Nevertheless, circumstances may change, and perhaps the candidates along with them, to the benefit of the United States and the region. Public opinion may not remain marginalized and easily ignored. The concentrations of domestic economic power that largely shape policy may come to recognize that their interests are better served by joining the general public, and the rest of the world, than by accepting Washington's hard line.

NOTE

1. Adapted from *Perilous Power: The Middle East and U.S. Foreign Policy* by Noam Chomsky and Gilbert Achcar, updated paperback edition, Paradigm Publishers, September 2008. Reprinted with permission. Copyright © 2008 by Noam Chomsky and Gilbert Achcar.

Delaying Doomsday:
This Century's Challenges

APRIL 24, 2008

The primary challenge facing the people of the world is, literally, survival.

General Lee Butler, former head of the U.S Strategic Command (STRATCOM), put the matter plainly a decade ago. Throughout his long military career he was "among the most avid of these keepers of the faith in nuclear weapons," he wrote, but it is now his "burden to declare with all of the conviction I can muster that in my judgment they served us extremely ill."

Butler raises a haunting question: "By what authority do succeeding generations of leaders in the nuclear-weapons states usurp the power to dictate the odds of continued life on our planet? Most urgently, why does such breathtaking audacity persist at a moment when we should stand trembling in the face of our folly and united in our commitment to abolish its most deadly manifestations?"

To our shame, his question not only remains unanswered, even unasked, but also has taken on greater urgency.

Butler may have been reacting to one of the most astonishing planning documents in the available record, the 1995 report of STRATCOM, "Essentials of Post–Cold War Deterrence."

The report advised that the military resources directed against the former Soviet Union be maintained, but with an expanded mission. They must also be directed against "rogue states" of the Third World, in accord with the Pentagon view that "the international environment has now evolved from a 'weapon rich environment' [the Soviet Union] to a 'target rich environment' [the Third World]."

Even if not used, "nuclear weapons always cast a shadow over any crisis or conflict," STRATCOM observed, enabling us to gain our ends through intimidation.

Nuclear weapons "seem destined to be the centerpiece of U.S. strategic deterrence for the foreseeable future." We must reject a "no first use policy," and make it clear to adversaries that our "reaction" may "either be response or preemptive."

Furthermore, "it hurts to portray ourselves as too fully rational and cool-headed." The "national persona we project" should make clear "that the United States may become irrational and vindictive if its vital interests are attacked—and that "some elements may appear to be potentially 'out of control.'"

Apart from the dissident margins, the report appears to have elicited no interest.

Forty years earlier, Bertrand Russell and Albert Einstein had warned that we face a choice that is "stark and dreadful and inescapable: Shall we put an end to the human race; or shall mankind renounce war?" They were not exaggerating.

Environmental catastrophe is no less a threat to survival, in a not too distant future. A serious approach will surely require significant socioeconomic changes and dedication of resources to technological innovations, particularly harnessing solar energy, many scientists contend.

A related threat is limited access to the basic means

of life: water and sufficient food. Short-term solutions include desalination, for example, in which Saudi Arabia is well in the lead in scale, and Israel in technology. This is one of many bases for constructive cooperation—if the United States and Israel would permit a resolution of the Israel-Palestine conflict in terms of the international consensus on a viable two-state settlement that they have been barring for thirty years, with rare and brief departures, another critical challenge with broad repercussions.

There are many uncertainties about how to address these issues. We can, however, be confident that the longer the delay in confronting them, the greater will be the cost to coming generations.

At least it is clear how to end the threat of nuclear weapons: eliminate them, a legal obligation of the nuclear powers, as the World Court determined a decade ago.

More broadly, sensible plans exist to restrict all production of weapons-usable fissile materials to an international agency, to which states can apply for nonmilitary uses. The U.N. Committee on Disarmament has already voted for a verifiable treaty with these provisions, in November 2004. The vote was 147 to 1 (the United States) with two abstentions (Israel and Britain).

An important interim step would be the establishment of nuclear-weapons-free zones (NWFZs). A number of such zones already exist, for example in Africa, the South Pacific and Southeast Asia, though as always, their significance depends on the willingness of the great powers to observe the rules. In Africa and the South Pacific, the United States refuses to do so, maintaining nuclear weapons in bases it controls (Diego Garcia, Pacific Islands), and insisting on transit of nuclear weapons.

Nowhere would the establishment of nuclear-weapons-free zones be more valuable than in the Middle East. In April 1991, the U.N. Security Council affirmed "the goal of establishing in the Middle East a zone free from weapons of mass destruction and all missiles for their delivery and the objective of a global ban on chemical weapons" (Resolution 687, Article 14).

The commitment is of particular significance for the United States and United Kingdom, since it is that resolution on which they relied in seeking a thin legal justification for their invasion of Iraq.

The goal of a Middle East nuclear-weapons-free zone has been endorsed by Iran, and is supported by a large majority of Americans and Iranians. It is, however, dismissed by the U.S. government and both political parties, and is virtually unmentionable in mainstream discussion.

A large majority of Americans and Iranians, as well as the developing countries (the Group of 77, now with 130 members), also agree that Iran has the "inalienable rights" of all parties to the Nonproliferation Treaty (NPT) "to develop research, production and use of nuclear energy for peaceful purposes without discrimination," rights that would also extend to Israel, Pakistan and India, were they to accept the NPT.

When the press reports, as it commonly does, that Iran is defying "the world" by enriching uranium, it is adopting an interesting concept of "the world."

As for inspection, the International Atomic Energy Agency (IAEA) has proven to be highly competent, and with great-power support, could become more so.

Zeev Maoz, one of Israel's leading strategic analysts, has presented solid arguments that Israel's nuclear programs harm its security, and has urged Israel to "seriously reconsider its nuclear policy and explore using its nuclear leverage to bring about a regional agreement for a weapons-of-mass-destruction-free zone (WMDFZ) in the Middle East."

With its overwhelming power, Washington's stand is of course decisive. The Bush administration has been praised in the West for a perceived recent shift from aggressive militarism toward diplomacy, but the admiration is not universal.

Commenting on President Bush's January 2008 visit to the Gulf states, Middle East specialist and former Ambassador Charles Freeman writes that "Arabs are notoriously courteous and welcoming to guests, even when they don't like those guests. . . . Yet, when the American president visited and spoke on the subject of Iran, he drew an editorial in Saudi Arabia's major English language newspaper deploring the fact that 'American policy represents not diplomacy in search of peace, but madness in search of war.'"

Developments in Europe are also fraught with danger. To NATO leaders, it is the merest truism that they themselves are a force for peace. Most of the world, which has rather different memories of Western benevolence, sees matters differently. So does Russia.

There seemed to be hope for long-term peace in Europe when the Soviet Union collapsed. Mikhail Gorbachev agreed to allow a unified Germany to join NATO, an astonishing concession in the light of history; Germany alone had practically destroyed Russia twice in that century, and now would belong to a hostile military alliance led by the global superpower.

There was a quid pro quo: President Bush I agreed

that NATO would not expand to the East, granting Russia some measure of security. In violation of a verbal agreement, NATO expanded at once to East Germany. President Bill Clinton reneged further on the agreement. NATO not only expanded eastward but also rejected the proposal of Russia (with Ukraine and Belarus) to establish a formal nuclear-weapons-free zone from the Arctic to the Black Sea, encompassing central Europe.

In response, Russia withdrew the policy of no-first-use of nuclear weapons that it had adopted after the Bush-Gorbachev agreement, reverting to the first-use policy that NATO has never abandoned.

Tensions mounted rapidly when Bush II took office, with threatening rhetoric, sharp expansion of offensive military capacity, withdrawal from key security treaties, and direct aggression. As predicted, Russia responded by increasing its own military capacity, followed later by China.

Ballistic missile defense (BMD) programs are a particular threat. Ballistic missile defense is understood on all sides to be a first-strike weapon, perhaps capable of nullifying a retaliatory strike and thus undermining deterrent capacity. The quasi-governmental Rand corporation describes BMD as "not simply a shield but an enabler of U.S. action."

In journals across the political spectrum, military analysts write approvingly of ballistic missile defense. In the conservative *National Interest*, Andrew Bacevich writes, "Missile defense isn't really meant to protect America. It's a tool for global dominance." To Lawrence Kaplan in the liberal *New Republic*, ballistic missile defense is "about preserving America's ability to wield power abroad. It's not about defense. It's about offense. And that's exactly why we need it."

Russian strategists draw the same conclusion. They can hardly fail to regard U.S. BMD installations in northern Poland and the Czech Republic as serious potential threats to their security, conclude U.S. analysts George Louis and Theodore Postol.

U.S. policymakers have long taken for granted the right to dominate the world. But the area of U.S. dominance is eroding, even at the core.

In recent years, South America has been taking steps to escape U.S. control. The countries of the region are moving toward integration, a prerequisite for independence, while also addressing severe internal problems, most importantly, the traditional rule of a rich and often white minority over a sea of misery and suffering.

South-South relations are also strengthening, linking Brazil, South Africa and India, among other interactions. And as in Africa and the Middle East, the rising economic power of China is providing alternatives to Western dominance.

For some years, the international economy has been tripolar, with major centers in North America, Europe and East/Northeast Asia, now increasingly South Asia and Southeast Asia as well.

The United States reigns supreme in one dimension: means of violence, in which it spends roughly as much as the rest of the world combined, and is technologically far more advanced. But in other respects the world is becoming more diverse and complex.

The two traditional modes of U.S. control are violence and economic strangulation. They may be losing their efficacy, but by no means have they been abandoned.

In March 2008, the U.S. Treasury Department warned the world's financial institutions against dealings with Iran's major state-owned banks. The warnings have teeth, thanks to a provision of the USA PATRIOT Act that enables Washington to bar any financial institution violating U.S. directives from access to the U.S. financial system.

That is a threat that few will dare to face, possibly even China. Economic analyst John McGlynn hardly exaggerates when he describes the March [2008] Treasury warning as a declaration of war against Iran, which might substantially isolate Iran from the international economy.

McGlynn's analysis receives support from an unexpected source: a manifesto for a militant "new grand strategy" issued in January 2008 by five high-level former NATO commanders, advising that "nuclear weapons—and with them the option of first use—are indispensable, since there is simply no realistic prospect of a nuclear-free world."

Among the potential "acts of war" that we must guard against, the commanders include "abusing the leverage" provided by "weapons of finance." To be sure, they adopt the conventional doctrine that brandishing such weapons becomes an "act of war" only when they are in the hands of others. When we use them—actually, not potentially—they are righteous means of self-defense, as are any aggressive actions by favored states, throughout history.

It is a marvel that the species has survived this long in an era of nuclear weapons. No one should take lightly the warning of "Apocalypse soon," in Robert McNamara's words, if we pursue our present course.

Failure to confront the challenges that lie ahead may well confirm the speculation of one of the great figures in modern biology, Ernst Mayr, that higher intelligence is an evolutionary error, incapable of survival for more than a passing moment of evolutionary time.

Middle East Road Trip

MAY 29, 2008

In mid-May [2008], President Bush traveled to the Middle East to establish his legacy more firmly in the part of the world that has been the prime focus of his presidency.

The trip had two principal destinations, each chosen to celebrate a major anniversary: Israel, the sixtieth anniversary of its founding and recognition by the United States; and Saudi Arabia, the seventy-fifth anniversary of U.S. recognition of the newly founded kingdom.

The choices made good sense in the light of history and the enduring character of U.S. Middle East policy: control of oil, and support of the proxies who help maintain it.

An omission, however, was not lost on the people of the region. Though Bush celebrated the founding of Israel, he did not recognize (let alone commemorate) the paired event from sixty years ago: the destruction of Palestine, the *Nakba*, as Palestinians refer to the events that expelled them from their lands.

During his three days in Jerusalem, the president was an enthusiastic participant in lavish events and made sure to go to Masada, a near-sacred site of Jewish nationalism.

But he did not visit the seat of the Palestinian authority in Ramallah, or Gaza City, or a refugee camp, or the town of Qalqilya—strangled by the Separation Wall, now becoming an Annexation Wall under the illegal Israeli settlement and development programs that Bush has endorsed officially, the first president to do so.

And it was out of the question that he would have any contact with Hamas leaders and parliamentarians, chosen in the only free election in the Arab world, many of them in Israeli jails with no pretense of judicial proceedings.

The pretexts for this stance scarcely withstand a moment's analysis. Also of no moment is the fact that Hamas has repeatedly called for a two-state settlement in accord with the international consensus that the United States and Israel have rejected, virtually alone, for more than thirty years, and still do.

Bush did allow the U.S. favorite, Palestinian president Mahmoud Abbas, to participate in meetings in Egypt with many regional leaders.

Bush's last visit to Saudi Arabia was in January [2008]. On both trips, he sought, without success, to draw the kingdom into the anti-Iranian alliance he has been seeking to forge. That is no small task, despite the concern of the Sunni rulers over the "Shiite crescent" and growing Iranian influence, regularly termed "aggressiveness."

For the Saudi rulers, accommodation with Iran may be preferable to confrontation. And though public opinion is marginalized, it cannot be completely dismissed. In a recent poll of Saudis, Bush ranked far above Osama bin Laden in the "very unfavorable" category, and more than twice as high as Iranian President Ahmadinejad and Hassan Nasrallah, leader of Hezbollah, Iran's Shiite ally in Lebanon.

U.S.-Saudi relations date to the recognition of the Kingdom in 1933—not coincidentally, the year when Standard of California obtained a petroleum concession and American geologists began to explore what turned out to be the world's largest reserves of oil.

The United States quickly moved to ensure its own control, important steps in a process by which the United States took over world dominance from Britain, which was slowly reduced to a "junior partner," as the British Foreign Office lamented, unable to counter "the economic imperialism of American business interests, which is quite active under the cloak of a benevolent and avuncular internationalism" and is "attempting to elbow us out."

The strong U.S.-Israel alliance took its present form in 1967, when Israel performed a major service to the United States by destroying the main center of secular Arab nationalism, Nasser's Egypt, also safeguarding the Saudi rulers from the secular nationalist threat. U.S. planners had recognized a decade earlier that a "logical corollary" of U.S. opposition to "radical" (that is, independent) Arab nationalism would be "to support Israel as the only strong pro-Western power left in the Middle East."

Investment by U.S. corporations in Israeli high-tech industry has sharply increased, including Intel, Hewlett Packard, Microsoft, Warren Buffett and others, joined by major investors from Japan and India—in the latter case, one facet of a growing U.S.-Israel-India strategic alliance.

To be sure, other factors underlie the U.S.-Israeli relationship. In Jerusalem, Bush invoked "the bonds of the book," the faith "shared by Christians like himself and Jews," the Australian press reported, but apparently not shared by Muslims or even Christian Arabs, like those in Bethlehem, now barred from Jerusalem, a few kilometers away, by illegal Israeli construction projects.

The *Saudi Gazette* bitterly condemned Bush's "audacity

to call Israel the 'homeland for the chosen people'"—the terminology of ultra-religious Israeli hard-liners. The *Gazette* added that Bush's "particular brand of moral bankruptcy was on full display when he made only passing mention of a Palestinian state in his vision of the region 60 years hence."

It is not difficult to discern why Bush's chosen legacy should stress relations with Israel and the rulers of Saudi Arabia, with a side glance at Egypt, along with disdain for the Palestinians and their miserable plight, apart from a few ritual phrases.

We need not tarry on the thought that the president's choices have anything to do with justice, human rights or the vision of "democracy promotion" that gripped his soul as soon as the pretexts for the invasion of Iraq had collapsed.

But the choices do accord with a general principle, observed with considerable consistency: Rights are assigned in accord with service to power.

Palestinians are poor, weak, dispersed and friendless. It is elementary, then, that they should have no rights. In sharp contrast, Saudi Arabia has incomparable resources of energy, Egypt is the major Arab state, and Israel is a rich Western country and the regional powerhouse, with air and armored forces that are reported to be larger and technologically more advanced than those of any NATO power (apart from its patron) along with hundreds of nuclear weapons, and with an advanced and largely militarized economy closely linked to the United States.

The contours of the intended legacy are therefore quite predictable.

Iraq Oil: A Deal With The Devil

JULY 6, 2007

The deal just taking shape between Iraq's Oil Ministry and
four Western oil companies raises critical questions about
the nature of the U.S. invasion and occupation of Iraq—
questions that should certainly be addressed by presidential candidates and seriously discussed in the United States,
and of course in occupied Iraq, where it appears that the
population has little if any role in determining the future
of their country.

Negotiations are under way for Exxon Mobil, Shell,
Total and BP—the original partners decades ago in the
Iraq Petroleum Company, now joined by Chevron and
other smaller oil companies—to renew the oil concession they lost to nationalization during the years when the
oil producers took over their own resources. The no-bid
contracts, apparently written by the oil corporations with
the help of U.S. officials, prevailed over offers from more
than forty other companies, including companies in China,
India and Russia.

"There was suspicion among many in the Arab world
and among parts of the American public that the United
States had gone to war in Iraq precisely to secure the oil
wealth these contracts seek to extract," Andrew E. Kramer
wrote in the *New York Times*.

Furthermore, it may be that the military occupation has taken the initiative in restoring the hated Iraq

Petroleum Company, which, as Seamus Milne writes in the London *Guardian*, was imposed under British rule to "dine off Iraq's wealth in a famously exploitative deal."

Later reports speak of delays in the bidding. Much is happening in secrecy, and it would be no surprise if new scandals emerge.

Concern over control of Iraqi oil is not hard to understand. Iraq contains perhaps the second-largest oil reserves in the world, which are, furthermore, very cheap to extract: no permafrost or shale or tar sands or deep-sea drilling. For U.S. planners, it is a priority that Iraq remain under U.S. control, to the extent possible, as an obedient client state that will also house major U.S. military bases, right at the heart of the world's major energy reserves.

That these were the primary goals of the invasion was always clear enough through the haze of successive pretexts: weapons of mass destruction, Saddam's links with al-Qaida, democracy promotion and the war against terrorism, which, as predicted, sharply increased as a result of the invasion.

Last November [2007], the guiding concerns were made explicit when President Bush and Iraq's Prime Minister, Nouri al-Maliki, signed a "Declaration of Principles," ignoring the U.S. Congress and Iraqi parliament, and the populations of the two countries.

The Declaration left open the possibility of an indefinite long-term U.S. military presence in Iraq that would presumably include the huge air bases now being built around the country, and the enormous "embassy" in Baghdad.

The Declaration also had a remarkably brazen statement about exploiting the resources of Iraq. It said that the economy of Iraq, which means its oil resources, must be open to foreign investment, "especially American investments." That comes close to a pronouncement that we invaded you so that we can control your country and have privileged access to your resources.

The seriousness of this commitment was underscored in January [2008], when President Bush issued a "signing statement" declaring that he would reject any congressional legislation that restricted funding "to establish any military installation or base for the purpose of providing for the permanent stationing of United States Armed Forces in Iraq" or "to exercise United States control of the oil resources of Iraq." There could hardly be a clearer statement of the real reasons for the invasion, hardly a surprise to those familiar with policy formation over many years.

Extensive resort to "signing statements" to expand executive power is yet another Bush innovation, condemned by the American Bar Association as "contrary to the rule of law and our constitutional separation of powers." To no avail.

Not surprisingly, the Declaration aroused immediate objections in Iraq, among others from Iraqi unions, which survive even under the harsh anti-labor laws that Saddam instituted and the occupation preserves.

In Washington propaganda, the spoiler to U.S. domination in Iraq is Iran. U.S. problems in Iraq are blamed on Iran. U.S. Secretary of State Condoleezza Rice sees a simple solution: "Foreign forces" and "foreign arms" should be withdrawn from Iraq—Iran's, of course not referring to ours.

The confrontation over Iran's nuclear programs heightens the tensions. The Bush administration's "regime change" policy toward Iran comes with ominous threats of force (there Bush is joined by both U.S. presidential candidates). The policy also is reported to include terrorism within Iran—again legitimate, for the world rulers. A majority of the American people favor diplomacy and oppose the use of force. But public opinion is largely irrelevant to policy formation, not just in this case.

An irony is that Iraq is turning into a U.S.-Iranian condominium. The Maliki government is the sector of Iraqi society most supported by Iran. The so-called Iraqi army— just another militia—is largely based on the Badr brigade, which was trained in Iran, and fought on the Iranian side during the Iran-Iraq war.

Nir Rosen, one of the most astute and knowledgeable correspondents in the region, observes that the main target of the U.S.-Maliki military operations, Moktada al-Sadr, is disliked by Iran as well: He's independent and has popular support, therefore he's dangerous.

Iran "clearly supported Prime Minister Maliki and the Iraqi government against what they described as 'illegal armed groups' (of Moktada's Mahdi army) in the recent conflict in Basra," Rosen writes, "which is not surprising given that their main proxy in Iraq, the Supreme Iraqi Islamic Council dominates the Iraqi state and is Maliki's main backer."

"There is no proxy war in Iraq," Rosen concludes, "because the U.S. and Iran share the same proxy."

Tehran is presumably pleased to see the United States institute and sustain a government in Iraq that's receptive to its influence. For the Iraqi people, however, that government continues to be a disaster, very likely with worse to come.

In *Foreign Affairs*, Steven Simon points out that current U.S. counterinsurgency strategy is "stoking the three forces that have traditionally threatened the stability of Middle Eastern states: tribalism, warlordism and sectarianism." The outcome might be "a strong, centralized state ruled by a military junta that would resemble" Saddam's regime.

If Washington achieves its goals, then its actions are taken to be justified. Reactions are quite different when Vladimir Putin succeeds in pacifying Chechnya, to an extent well beyond what General David Petraeus has achieved in Iraq. But that is THEM, and this is US. Criteria are therefore entirely different.

In the United States the Democrats are silenced now because of the supposed success of the U.S. military surge in Iraq. Their silence reflects the fact that there are no principled criticisms of the war. In this way of regarding the world, if you're achieving your goals, the war and occupation are justified. Sweetheart oil deals, if attainable, come with the territory.

In fact, the whole invasion is a war crime—indeed the supreme international crime, differing from other war crimes in that it encompasses all the evil that follows, in the terms of the Nuremberg judgment. This is among the topics that can't be discussed, in the presidential campaign or elsewhere. Why are we in Iraq? What do we owe Iraqis for destroying their country? The majority of the American people favor U.S. withdrawal from Iraq. Do their voices matter?

Nuclear Threats:
All Options Are on the Table

JULY 31, 2008

Nuclear threats and counter-threats are a subtext of our times, steadily, it seems, becoming more insistent.

The July [2008] meeting in Geneva between Iran and six major world powers on Iran's nuclear program ended with no progress. The Bush administration was widely praised for having shifted to a more conciliatory stand—namely, by allowing a U.S. diplomat to attend without participating—while Iran was castigated for failing to negotiate seriously. And the powers warned Iran that it would soon face more severe sanctions unless it terminated its uranium-enrichment programs.

Meanwhile India was applauded for agreeing to a nuclear pact with the United States that would effectively authorize its development of nuclear weapons outside the bounds of the Nonproliferation Treaty (NPT), with U.S. assistance in nuclear programs along with other rewards—in particular, to U.S. firms eager to enter the Indian market for nuclear and weapons development, and ample payoffs to parliamentarians who signed on, a tribute to India's flourishing democracy. Michael Krepon, co-founder of the Stimson Center and a leading specialist on nuclear threats, observed reasonably that Washington's decision in this case to "place profits ahead of nonproliferation" could mean the end of the NPT if others follow its lead, sharply increasing the dangers all around.

During the same period, Israel, another state that has defied the Nonproliferation Treaty with Western support, conducted large-scale military maneuvers in the eastern Mediterranean that were understood to be preparation for bombing Iran's nuclear facilities.

In a *New York Times* op-ed article, "Using Bombs to Stave Off War," the prominent Israeli historian Benny Morris wrote that Iran's leaders should welcome Israeli bombing with conventional weapons, because "the alternative is an Iran turned into a nuclear wasteland."

Purposely or not, Morris is reviving an old theme. During the 1950s, leading figures of Israel's governing Labor Party advised in internal discussion that "we will go crazy ("nishtagea") if crossed, threatening to bring down the Temple Walls in the manner of the first "suicide bomber," the revered Samson, who killed more Philistines by his suicide than in his entire lifetime.

Israel's nuclear weapons may well harm its own security, as Israeli strategic analyst Zeev Maoz persuasively argues. But security is often not a high priority for state planners, as history makes clear. And the "Samson complex," as Israeli commentators have called it, can be flaunted to warn the master to carry out the desired task of smashing Iran, or else we'll inflame the region and maybe the world.

The "Samson complex," reinforced by the doctrine that "the whole world is against us," cannot be lightly ignored. Shortly after the 1982 invasion of Lebanon, which left some 15,000 to 20,000 killed in an unprovoked effort to secure Israel's control of the occupied territories, Aryeh

Eliav, one of Israel's best-known doves, wrote that the attitude of "those who brought the 'Samson complex' here, according to which we shall kill and bury all the Gentiles around us while we ourselves shall die with them," is a form of "insanity" that was then all too prevalent, and still is.

U.S. military analysts have recognized that. As Army Lieutenant Colonel Warner Farr wrote in 1999, one "purpose of Israeli nuclear weapons, not often stated, but obvious, is their 'use' on the United States," presumably to ensure consistent U.S. support for Israeli policies—or else.

Others see further dangers. General Lee Butler, former commander in chief of the U.S. Strategic Command, observed in 1999 that "it is dangerous in the extreme that in the cauldron of animosities that we call the Middle East, one nation has armed itself, ostensibly, with stockpiles of nuclear weapons, perhaps numbering in the hundreds, and that inspires other nations to do so." This fact is hardly irrelevant to concerns about Iran's nuclear programs, but is off the agenda.

Also off the agenda is Article 2 of the U.N. Charter, which bars the threat of force in international affairs. Both U.S. political parties insistently proclaim their criminality, declaring that "all options are on the table" with regard to Iran's nuclear programs.

Some go beyond, like John McCain, joking about what fun it would be to bomb Iran and to kill Iranians, though the humor may be lost on the "unpeople" of the world, to borrow the term used by British historian Mark Curtis for those who do not merit the attention of the privileged and powerful.

Barack Obama declares that he would do "everything

in my power" to prevent Iran from gaining the capacity to produce nuclear weapons. The unpeople surely understand that launching a nuclear war would be in his power.

The chorus of denunciations of the New Hitlers in Tehran and the threat they pose to survival has been marred by a few voices from the back rooms. Former Mossad Chief Ephraim Halevy recently warned that an Israeli attack on Iran "could have an impact on us for the next 100 years."

An unnamed former senior Mossad official added, "Iran's achievement is creating an image of itself as a scary superpower when it's really a paper tiger"—which is not quite accurate: The achievement should be credited to U.S.-Israeli propaganda.

One of the participants in the July [2008] meetings was Egyptian Foreign Minister Ahmed Aboul Gheit, who outlined "the Arab position": "to work toward a political and diplomatic settlement under which Iran will maintain the right to use nuclear energy for peaceful purposes" but without nuclear weapons.

The "Arab position" is that of most Iranians, along with other unpeople. On July 30 [2008], the 120-member Nonaligned Movement reiterated its previous endorsement of Iran's right to enrich uranium in accord with the NPT.

Joining the unpeople is the large majority of Americans, according to polls. The American unpeople not only endorse Iran's right to enrich uranium for peaceful purposes but also support the "Arab position" calling for a nuclear-weapons-free zone in the entire region, a step that would sharply reduce major threats, but is also off the agenda of the powerful; unmentionable in electoral campaigns, for example.

Benny Morris assures us that "Every intelligence

agency in the world believes the Iranian program is geared toward making weapons." As is well known, the U.S. National Intelligence Estimate of November 2007 judged "with high confidence that in fall 2003, Tehran halted its nuclear weapons program." It is doubtful, to say the least, that the intelligence agencies of every country of the Non-aligned Movement disagree.

Morris is presumably reporting information from an Israeli intelligence source—which generalizes to "every intelligence agency" by the same logic that instructs us that Iran is defying "the world" by seeking to enrich uranium: the world apart from its unpeople.

There are rumblings in radical nationalist (so-called "neo-con") circles that if Barack Obama wins the election, Bush-Cheney should bomb Iran, since the threat of Iran is too great to be left in the hands of a wimpish Democrat. Reports also have surfaced—recently from Seymour Hersh in the *New Yorker*—on U.S. "covert operations" in Iran, known as international terrorism if conducted by enemies.

In June (2008), Congress came close to passing a resolution (H. Con. Res. 362), strongly supported by the Israeli lobby, virtually calling for a blockade of Iran—an act of war, that could have set off the conflagration that is greatly feared in the region and around the world. Pressures from the anti-war movement appear to have beaten back this particular effort, according to Mark Weisbrot at *Alternet .org*, but others are likely to follow.

The government of Iran merits severe condemnation on many counts, but the Iranian threat remains a desperate construction of those who arrogate to themselves the right to rule the world, and consider any impediment to their just rule to be criminal aggression. That is the primary threat that should concern us, as it concerns saner minds in the West, and the unpeople of the rest of the world.

Georgia and the Neo-con Cold Warriors

SEPTEMBER 9, 2008

Aghast at the atrocities by U.S. forces invading the Philippines, and the rhetorical flights about liberation and noble intent that routinely accompany crimes of state, Mark Twain threw up his hands at his inability to wield his formidable weapon of satire.

The immediate object of his frustration was the renowned General Frederick Funston. "No satire of Funston could reach perfection," Twain lamented, "because Funston occupies that summit himself . . . (he is) satire incarnated."

Twain's conceit often comes to mind, again in recent weeks during the Russia-Georgia-Ossetia war.

George Bush, Condoleezza Rice and other dignitaries solemnly invoked the sanctity of the United Nations and international law, warning that Russia could be excluded from international institutions "by taking actions in Georgia that are inconsistent with" U.N. principles.

The sovereignty and territorial integrity of all nations must be rigorously honored, they intoned—"all nations" apart from those that the United States chooses to attack: Iraq, Serbia, perhaps Iran, and a long, familiar list of others.

The junior partner joined in as well. British foreign secretary David Miliband accused Russia of engaging in "nineteenth-century forms of diplomacy" by invading a sovereign state, something Britain would never contemplate today.

Such an act "is simply not the way that international relations can be run in the 21st century," Miliband added, echoing the decider-in-chief, who said that invasion of "a sovereign neighboring state . . . is unacceptable in the 21st century."

The interplay of satire and real-life events becomes "even more enlightening," Serge Halimi wrote in *Le Monde Diplomatique*, "when, to defend his country's borders, the charming pro-American [Mikheil] Saakashvili repatriates some of the two thousand soldiers he had sent to invade Iraq," one of the largest contingents apart from the two warrior states.

Prominent analysts joined the chorus. Fareed Zakaria applauded Bush's observation that Russia's behavior is unacceptable today, unlike the nineteenth century, "when the Russian intervention would have been standard operating procedure for a great power." We therefore must devise a strategy for bringing Russia "in line with the civilized world," to whom intervention is unthinkable.

The seven charter members of the Group of Eight industrialized countries issued a statement "condemning the action of our fellow G8 member," Russia, which has yet to comprehend the long-standing Anglo-American commitment to nonintervention. The European Union held a rare emergency meeting to condemn Russia's crime, its first meeting since the invasion of Iraq, which elicited no condemnation.

Russia called for an emergency session of the U.N. Security Council, but no consensus was reached because, according to Council diplomats, the United States, Britain and some others rejected a phrase that called on both sides "to renounce the use of force."

The reactions recall Orwell's observations on the "indifference to reality" of the nationalist, who "not only does not disapprove of atrocities committed by his own side, but . . . has a remarkable capacity for not even hearing about them."

The basic history is not seriously in dispute. South Ossetia and Abkhazia (with its ports on the Black Sea) were assigned by Stalin to his native Georgia. (Now Western leaders sternly admonish that Stalin's directives must be respected.)

The provinces enjoyed relative autonomy until the collapse of the Soviet Union. In 1990, Georgia's ultranationalist president Zviad Gamsakhurdia abolished autonomous regions and invaded South Ossetia. The bitter war that followed left one thousand dead and tens of thousands of refugees.

A small Russian force supervised a long, uneasy truce, broken on August 7, 2008, when Georgian president Saakashvili ordered his forces to invade. According to "an extensive set of witnesses," the *New York Times* reports, Georgia's military at once "began pounding civilian sections of the city of Tskhinvali, as well as a Russian peacekeeping base there, with heavy barrages of rocket and artillery fire."

The predictable Russian response drove Georgian forces out of South Ossetia, and Russia went on to conquer parts of Georgia, then partially withdrawing to the vicinity of South Ossetia. There were many casualties and atrocities. As is normal, the innocent suffered severely.

In the background of the Caucasus tragedy lie two crucial issues. One is control over natural gas and oil pipelines from Azerbaijan to the West. Georgia was chosen by Bill Clinton to bypass Russia and Iran, and was also heavily militarized for the purpose. Hence Georgia is "a very major and strategic asset to us," Zbigniew Brzezinski observes.

It is noteworthy that analysts are becoming less reticent in explaining real U.S. motives in the region as pretexts of dire threats and liberation fade and it becomes more difficult to deflect Iraqi demands for withdrawal of the occupying army. Thus the editors of the *Washington Post* admonished Barack Obama for regarding Afghanistan as "the central front" for the United States, reminding him that Iraq "lies at the geopolitical center of the Middle East and contains some of the world's largest oil reserves," and Afghanistan's "strategic importance pales beside that of Iraq." A welcome, if belated, recognition of reality about the U.S. invasion.

The second divisive issue in the Caucasus is expansion of NATO to the East. As the Soviet Union collapsed, Mikhail Gorbachev made a concession that was astonishing in the light of recent history and strategic realities: He agreed to allow a united Germany to join a hostile military alliance.

Gorbachev agreed to the concession on the basis of "assurances that NATO would not extend its jurisdiction to the east, 'not one inch' in [Secretary of State] Jim Baker's exact words," according to Jack Matlock, the U.S. ambassador to Russia in the crucial years 1987 to 1991.

Bush-Baker and Clinton quickly reneged on that commitment, also dismissing Gorbachev's effort to end the

Cold War with cooperation among partners. And NATO rejected a Russian proposal for a nuclear-weapons-free-zone from the Arctic to the Black Sea, which would have "interfered with plans to extend NATO," strategic analyst and former NATO planner Michael McGwire observes.

Gorbachev's hopes were abandoned in favor of U.S. triumphalism. Clinton's steps were sharply escalated by Bush II's aggressive posture and actions. Matlock writes that Russia might have tolerated incorporation of former Russian satellites into NATO if the United States "had not bombed Serbia and continued expanding. But, in the final analysis, ABM missiles in Poland, and the drive for Georgia and Ukraine in NATO crossed absolute red lines. The insistence on recognizing Kosovo independence was sort of the very last straw. Putin had learned that concessions to the U.S. were not reciprocated, but used to promote U.S. dominance in the world. Once he had the strength to resist, he did so," in Georgia.

There is much talk about a "new cold war" instigated by brutal Russian behavior in Georgia. One cannot fail to be alarmed by new U.S. naval contingents in the Black Sea—the counterpart would hardly be tolerated in the Gulf of Mexico—and other signs of confrontation. Efforts to expand NATO to Ukraine, now contemplated, could become extremely hazardous. Vice President Cheney's recent visits to Georgia and Ukraine are recklessly provocative.

Nonetheless, a new Cold War seems unlikely. To evaluate the prospect, we should begin by clarity about the old Cold War. Fevered rhetoric aside, in practice the Cold War was a tacit compact in which each of the contestants was largely free to resort to violence and subversion to control its own domains: for Russia, its Eastern neighbors; for the global superpower, most of the world. Human society need not endure—and might not survive—a resurrection of anything like that.

A sensible alternative is the Gorbachev vision rejected by Clinton and undermined by Bush. Sane advice along these lines has recently been given by former Israeli Foreign Minister and historian Shlomo ben-Ami, writing in the Lebanese press:

"Russia must seek genuine strategic partnership with the U.S., and the latter must understand that, when excluded and despised, Russia can be a major global spoiler. Ignored and humiliated by the U.S. since the Cold War ended, Russia needs integration into a new global order that respects its interests as a resurgent power, not an anti-Western strategy of confrontation."

The Campaign and the Financial Crisis
OCTOBER 5, 2008

The simultaneous unfolding of the U.S. presidential campaign and unraveling of the financial markets presents one of those occasions where the political and economic systems starkly reveal their nature.

Passion about the campaign may not be universally shared, but almost everybody can feel the anxiety from the foreclosure of a million homes, and concerns about jobs, savings and health care at risk.

The initial Bush proposals to deal with the crisis were quickly modified. Under intense lobbyist pressure, they were reshaped as "a clear win for the largest institutions in the system . . . a way of dumping assets without having to fail or close," as described by James G. Rickards, who negotiated the federal bailout for the hedge fund Long Term Capital Management in 1998, reminding us that we are treading familiar turf.

The immediate origins of the current meltdown lie in the collapse of the housing bubble supervised by Federal Reserve Chairman Alan Greenspan, which sustained the struggling economy through the Bush years by debt-based consumer spending along with borrowing from abroad.

But the roots are deeper. In part they lie in the triumph of financial liberalization in the past thirty years—that is, freeing the markets as much as possible from government regulation. These steps predictably increased the frequency and depth of severe reversals, which now threaten to bring about the worst crisis since the Great Depression. Also predictably, the narrow sectors that reaped enormous profits from liberalization are calling for massive state intervention to rescue collapsing financial institutions.

Such interventionism is a regular feature of state capitalism, though the scale today is unusual. A study by international economists Winfried Ruigrok and Rob van Tulder fifteen years ago found that at least twenty companies in the Fortune 100 would not have survived if they had not been saved by their respective governments, and that many of the rest gained substantially by demanding that governments "socialize their losses," as in today's taxpayer-financed bailout. Such government intervention "has been the rule rather than the exception over the past two centuries," they conclude.

In a functioning democratic society, a political campaign would address such fundamental issues, looking into root causes and cures, and proposing the means by which people suffering the consequences can take effective control.

The financial market "underprices risk" and is "systematically inefficient," as economists John Eatwell and Lance Taylor wrote a decade ago, warning of the extreme dangers of financial liberalization and reviewing the substantial costs already incurred—and proposing solutions, which have been ignored. One factor is failure to calculate the costs to those who do not participate in transactions. These "externalities" can be huge. Ignoring systemic risk leads to more risk-taking than would take place in an efficient economy, even by the narrowest measures.

The task of financial institutions is to take risks, and if

well managed, to ensure that potential losses to themselves will be covered. The emphasis is on "to themselves." Under state capitalist rules, it is not their business to consider the cost to others—the "externalities" of decent survival—if their practices lead to financial crisis, as they regularly do.

Financial liberalization has effects well beyond the economy. It has long been understood that it is a powerful weapon against democracy. Free capital movement creates what some economists have called a "virtual parliament" of investors and lenders, who closely monitor government programs and "vote" against them if they are considered irrational: for the benefit of people, rather than concentrated private power.

Investors and lenders can "vote" by capital flight, attacks on currencies and other devices offered by financial liberalization. That is one reason why the Bretton Woods system established by the United States and Britain after World War II instituted regulated currencies and permitted capital controls.

The Great Depression and the war had aroused powerful radical democratic currents, ranging from the antifascist resistance to working-class organization. These pressures made it necessary to permit social democratic policies. The Bretton Woods system was designed in part to create a space for government action responding to public will—for some measure of democracy, that is.

John Maynard Keynes, the British negotiator, considered the most important achievement of Bretton Woods to be establishment of the right of governments to restrict capital movement. In dramatic contrast, in the neoliberal phase after the breakdown of the Bretton Woods system in the 1970s, the U.S. Treasury now regards free capital mobility as a "fundamental right," unlike such alleged "rights" as those guaranteed by the Universal Declaration of Human Rights: health, education, decent employment, security and other rights that the Reagan and Bush administrations have dismissed as "letters to Santa Claus," "preposterous," mere "myths."

In earlier years the public had not been much of a problem. The reasons are reviewed by Barry Eichengreen in his standard scholarly history of the international monetary system. He explains that in the nineteenth century, governments had not yet been "politicized by universal male suffrage and the rise of trade unionism and parliamentary labor parties." Therefore the severe costs imposed by the virtual parliament could be transferred to the general population.

But with the radicalization of the general public during the Great Depression and the anti-fascist war, that luxury was no longer available to private power and wealth. Hence in the Bretton Woods system, "limits on capital mobility substituted for limits on democracy as a source of insulation from market pressures."

The obvious corollary is that after the dismantling of the postwar system in the 1970s, democracy is restricted. It has therefore become necessary to control and marginalize the public in some fashion, processes particularly evident in the more business-run societies like the United States. The management of electoral extravaganzas by the public relations industry is one illustration.

"Politics is the shadow cast on society by big business," concluded America's leading twentieth-century social philosopher, John Dewey, and will remain so as long as power resides in "business for private profit through private control of banking, land, industry, reinforced by command of the press, press agents and other means of publicity and propaganda."

The United States effectively has a one-party system, the business party, with two factions, Republicans and Democrats. There are differences between them. In his study *Unequal Democracy: The Political Economy of the New Gilded Age*, Larry M. Bartels shows that during the past six decades "real incomes of middle-class families have grown twice as fast under Democrats as they have under Republicans, while the real incomes of working-poor families have grown six times as fast under Democrats as they have under Republicans."

Differences can be detected in the current election as well. Voters should consider them, but without illusions about the political parties, and with the recognition that consistently over the centuries, progressive legislation and social welfare have been won by popular struggles, not gifts from above.

Those struggles follow a cycle of success and setback. They must be waged every day, not just once every four years, always with the goal of creating a genuinely responsive democratic society, from the voting booth to the workplace.

Challenges for Barack Obama: Part 1
The Election and the Economy
NOVEMBER 25, 2008

The word that immediately rolled off of every tongue after the presidential election was "historic." And rightly so. A black family in the White House is truly a momentous event.

There were some surprises. One was that the election was not over after the Democratic convention. By usual indicators, the opposition party should have won by a landslide during a severe economic crisis, after eight years of disastrous policies on all fronts, including the worst record on job growth of any postwar president and a rare decline in median wealth, an incumbent so unpopular that his own party had to disavow him, and a dramatic collapse in U.S. standing in world opinion.

As many studies show, both parties are well to the right of the population on many major issues, domestic and international. Perhaps neither party reflects public opinion at a time when 80 percent of Americans think that the country is going in the wrong direction and that the government is run by "a few big interests looking out for themselves," not for the people, and a stunning 94 percent object that government does not attend to public opinion.

It could be argued that no party speaking for the public would be viable in a society that is business-run to such an unusual extent. At a very general level, the public's disenfranchisement is illustrated by the predictive success of political economist Thomas Ferguson's "investment theory" of politics, which holds that policies tend to reflect the wishes of the powerful blocs that invest every four years to control the state.

In some ways the election followed familiar patterns. The John McCain campaign was honest enough to announce clearly that the election wouldn't be about issues. Obama's message of "hope" and "change" offered a blank slate on which supporters could write their wishes. One could search websites for position papers, but correlation of these to policies is hardly spectacular, and in any event, what enters into voters' choices is what the campaign places front and center, as party managers know well.

There Obama's campaign greatly impressed the public relations industry, which named him "Advertising Age's marketer of the year for 2008," easily beating out Apple. The advertising industry's prime task is to ensure that uninformed consumers make irrational choices, thus undermining market theories that are based on just the opposite. And public relations recognizes the benefits of undermining democracy the same way.

The Center for Responsive Politics reports that once again elections were bought: "The best-funded candidates won nine out of 10 contests, and all but a few members of Congress will be returning to Washington."

Before the conventions, the viable candidates with most funding from financial institutions were Obama and McCain, with 36 percent each. Preliminary results indicate that by the end, Obama's campaign contributions, by industry, were concentrated among law firms (including lobbyists) and financial institutions. The investment theory of politics suggests some conclusions about the guiding policies of the new administration.

The power of financial institutions reflects the increasing shift of the economy from production to finance since the liberalization of finance in the 1970s, a root cause of the current scourges: the financial crisis, recession in the real economy (that is, production and consumption of goods) and the consequences for the large majority of Americans, whose real wages have stagnated for thirty years, while benefits declined.

Putting aside soaring rhetoric about hope and change, what can we realistically expect of an Obama administration?

Obama's choice of staff sends a strong signal. The first selection was for vice president: Joe Biden, one of the strongest supporters of the Iraq invasion among Senate Democrats and a longtime Washington insider, who consistently votes with his fellow Democrats but not always, as when he supported a measure to make it harder for individuals to erase debt by declaring bankruptcy.

The first post-election appointment was for the crucial position of chief of staff: Rahm Emanuel, one of the strongest supporters of the Iraq invasion among House Democrats and, like Biden, a longtime Washington insider.

Emanuel is also one of the biggest recipients of Wall Street campaign contributions, the Center for Responsive Politics reports. He "was the top House recipient in the 2008 election cycle of contributions from hedge funds, private equity firms and the larger securities/investment industry."

Since being elected to Congress in 2002, Emanuel "has received more money from individuals and PACs in the securities and investment business than any other industry." His task is to oversee Obama's approach to the worst financial crisis since the 1930s, for which his and Obama's funders share ample responsibility.

In an interview with the editor of the *Wall Street Journal*, Emanuel was asked what the Obama administration would do about "the Democratic congressional leadership, which is brimming with left-wing barons who have their own agenda," such as slashing defense spending (in accord with the will of the majority of the population) and "angling for steep energy taxes to combat global warming."

"Barack Obama can stand up to them," Emanuel assured the editor. The administration will be "pragmatic," fending off left extremists.

Obama's transition team is headed by John Podesta, Bill Clinton's chief of staff. Two other Clinton veterans, Robert Rubin and Lawrence Summers, are among the leading figures in his economic team. Rubin and Summers were both enthusiasts for the deregulation that was a major factor in the current financial crisis. As Clinton's Treasury Secretary, Rubin worked hard to abolish the Glass-Steagall act, which had separated commercial banks from financial institutions that incur high risks.

Financial economist Tim Canova writes that Rubin had "a personal interest in the demise of Glass-Steagall." Soon after leaving his position as Treasury Secretary, Rubin became "chair of Citigroup, a financial-services conglomerate that was facing the possibility of having to sell off its insurance underwriting subsidiary . . . the Clinton administration never brought charges against him for his obvious violations of the Ethics in Government Act."

Rubin was replaced as Treasury Secretary by Summers,

who presided over legislation barring federal regulation of derivatives, the "weapons of mass destruction" (as Warren Buffett calls them) that helped plunge financial markets to disaster.

Summers ranks as "one of the main villains in the current economic crisis," according to Dean Baker, among the handful of economists who warned of the impending crisis. Placing financial policy in the hands of Rubin and Summers, Baker observes, is "a bit like turning to Osama Bin Laden for aid in the war on terrorism."

Now Rubin and Summers are calling for regulation to help clean up the chaos they helped create.

The business press reviewed the records of Obama's Transition Economic Advisory Board, which met on November 7 [2008] to determine how to deal with the financial crisis. In *Bloomberg News*, Jonathan Weil concluded that "Many of them should be getting subpoenas as material witnesses right about now, not places in Obama's inner circle." About half "have held fiduciary positions at companies that, to one degree or another, either fried their financial statements, helped send the world into an economic tailspin, or both." Is it really plausible that "they won't mistake the nation's needs for their own corporate interests?"

The primary concern for the administration will be to arrest the financial crisis and the simultaneous recession in the real economy. But there is also a monster in the closet: the notoriously inefficient privatized health care system, which threatens to overwhelm the federal budget if current tendencies persist.

A majority of the public has long favored a national health care system, which should be far less expensive and more effective, comparative evidence indicates (along with many studies). As recently as 2004, any government intervention in the health care system was described in the press as "politically impossible" and "lacking political support"—meaning: opposed by the insurance industry, pharmaceutical corporations and others who count.

In 2008, however, first John Edwards, then Barack Obama and Hillary Clinton, advanced proposals that approach what the public has long preferred. These ideas now have "political support." What has changed? Not public opinion, which remains much as before. But by 2008, major sectors of power, primarily manufacturing industry, had come to recognize that they are being severely damaged by the privatized health care system. Hence the public will is coming to have "political support."

There is a long way to go, but the shift tells us something about the dysfunctional democracy in which the new administration finds its way.

Challenges for Barack Obama: Part 2
Iraq, Pakistan and Afghanistan

DECEMBER 18, 2008

Barack Obama's willingness to "talk" with enemies became a defining issue during the campaign. Can he live up to that pledge?

Diplomacy is the only sane alternative to the cycle of violence from the Middle East to Central Asia that threatens to engulf the world. A corollary is to recognize that violence tends to beget violence. It would also help if the Obama administration, and the West, faced up to the unannounced issues that drive policy in the region.

IRAQ

The government of Iraq has forged a Status of Forces Agreement (SOFA), reluctantly accepted by Washington, that is intended to end the U.S. military presence. The SOFA is the latest step in a process of mass nonviolent resistance that has compelled Washington, step by step, to agree to elections and increased independence of the occupied country.

An Iraqi spokesman said that the tentative SOFA "matches the vision of U.S. president-elect Barack Obama." Obama's "vision" isn't spelled out clearly, but he would probably go along in some fashion with the Iraqi government's demands. If so, that would require modification of officially announced U.S. plans to ensure control over Iraq's enormous oil resources while establishing bases to reinforce its dominance over the world's major energy-producing region.

It is noteworthy that recent worldwide polls show strong opposition to U.S. naval bases in the Gulf. Opposition is particularly strong within the region.

The prospect of shifting forces from Iraq to Afghanistan evoked a lesson from the editors of the *Washington Post*, quoted earlier: "While the United States has an interest in preventing the resurgence of the Afghan Taliban, the country's strategic importance pales beside that of Iraq, which lies at the geopolitical center of the Middle East and contains some of the world's largest oil reserves."

The NATO command is also coming to acknowledge publicly the crucial energy issues. In June 2007, NATO Secretary-General Jaap de Hoop Scheffer informed a meeting of members that "NATO troops have to guard pipelines that transport oil and gas that is directed for the West," and more generally to protect sea routes used by tankers and other "crucial infrastructure" of the energy system.

The task presumably includes the projected $7.6 billion TAPI pipeline that would deliver natural gas from Turkmenistan to Pakistan and India, running through Afghanistan's Kandahar province, where Canadian troops are deployed. The goal is "to block a competing pipeline that would bring gas to Pakistan and India from Iran" and to "diminish Russia's dominance of Central Asian energy exports," the Toronto *Globe and Mail* reported, plausibly outlining some of the contours of the new "Great Game" (when Britain and Russia vied for influence in Central Asia during the nineteenth century).

PAKISTAN

Obama has endorsed the Bush policy of attacking suspected al-Qaida leaders in countries that the United States has not (yet) invaded. In particular, he has not criticized the raids by Predator drones that have killed many civilians in Pakistan—and was soon to expand them radically as part of his global assassination campaign.

Right now a vicious mini-war is being waged in the tribal area of Bajaur in Pakistan, next to Afghanistan. The BBC describes widespread destruction from intense combat: "Many in Bajaur trace the roots of the uprising to a suspected U.S. missile strike on an Islamic seminary, or madrassa, in November 2006, which killed around 80 people."

The attack was reported in the mainstream Pakistani press by the highly respected dissident physicist Pervez Hoodbhoy but ignored in the United States. Things often look different at the other end of the club.

Hoodbhoy observed that the usual outcome of such attacks "has been flattened houses, dead and maimed children, and a growing local population that seeks revenge against Pakistan and the U.S." Bajaur today may illustrate the familiar cycle that Obama shows no sign of breaking.

On November 3 [2008], General David Petraeus, newly appointed head of the U.S. Central Command that covers the Middle East, had his first meeting with Pakistani President Asif Ali Zardari, army chief General Ashfaq Parvez Kayani and other officials.

Their primary concern: "The continuing drone attacks on our territory, which result in loss of precious lives and property, are counterproductive and difficult to explain by a democratically elected government," Zardari told Petraeus. His government, he said, is "under pressure to react more aggressively" to the strikes. These could lead to "a backlash against the U.S.," already deeply unpopular in Pakistan.

Petraeus said that he had heard the message, and that "we would have to take (Pakistani opinion) on board" when attacking the country—a practical necessity, no doubt, when more than 80 percent of the supplies for the U.S.-NATO war in Afghanistan pass through Pakistan.

How Pakistani opinion was "taken on board" was revealed two weeks later in the *Washington Post*, which reported that the United States and Pakistan reached "tacit agreement in September [2008] on a don't-ask-don't-tell policy that allows unmanned Predator aircraft to attack suspected terrorist targets" in Pakistan, according to unidentified senior officials in both countries. "The officials described the deal as one in which the U.S. government refuses to publicly acknowledge the attacks while Pakistan's government continues to complain noisily about the politically sensitive strikes."

The day before the report on the "tacit agreement" appeared, a suicide bombing in the conflicted tribal areas killed eight Pakistani soldiers—retaliation for an attack by a Predator drone that killed twenty people, including two Taliban leaders. The Pakistani parliament called for dialogue with the Taliban. Echoing the resolution, Pakistani foreign Minister Shah Mehmood Qureshi said, "There is an increasing realization that the use of force alone cannot yield the desired results."

AFGHANISTAN

Afghan President Hamid Karzai's first message to President-elect Obama was much like that delivered to Petraeus by Pakistani leaders: "End U.S. airstrikes that risk civilian casualties." His message was sent shortly after coalition troops bombed a wedding party in Kandahar province, reportedly killing forty people. There is no indication that his opinion was "taken on board."

The British command has warned that there is no military solution to the conflict in Afghanistan and that there will have to be negotiations with the Taliban, risking a rift with the United States, the *Financial Times* reports. Issues are already on the table, writes Jason Burke, a correspondent for the *Observer* with long experience in the region: "The Taliban have been engaged in secret talks about ending the conflict in Afghanistan in a wide-ranging 'peace process' sponsored by Saudi Arabia and supported by Britain."

Some Afghan peace activists have reservations about this approach, preferring a solution without foreign interference. A growing network of peace activists is calling for negotiations and reconciliation with the Taliban in a National Peace Jirga, a grand assembly of Afghans, formed in May 2008.

At a meeting in May [2008] in support of the Jirga, three thousand Afghan political and intellectuals, mainly Pashtuns, the largest ethnic group, criticized "the international military campaign against Islamic militants in Afghanistan and called for dialogue to end the fighting," Agence France-Presse reported.

The interim chairman of the National Peace Jirga, Bakhtar Aminzai, "told the opening gathering that the current conflict could not be resolved by military means and that only talks could bring a solution."

A leader of Awakened Youth of Afghanistan, a prominent anti-war group, said that we must end "Afghanicide—the killing of Afghanistan."

Polling in war-torn Afghanistan is difficult, but the results merit attention. A Canadian-run poll already mentioned found that Afghans favor the presence of Canadian and other foreign troops—with what mission is not clear—but few anticipate that "the Taliban will prevail once foreign troops leave," a large majority support a negotiated settlement and more than half favor a coalition government. It appears, then, that the great majority favor peaceful means.

A study of Taliban foot soldiers by the *Globe and Mail*, though not a scientific survey as the newspaper points out, nevertheless yields considerable insight. All were Afghan Pashtuns, from the Kandahar area. They described themselves as mujahadeen, following the ancient tradition of driving out foreign invaders. Almost a third reported that at least one family member had died in aerial bombings in recent years. Many said that they were fighting to defend Afghan villagers from air strikes by foreign troops. Few claimed to be fighting a global jihad, or had allegiance to Taliban leader Mullah Omar. Most saw themselves as fighting for principles—an Islamic government—not a leader.

Again, these results suggest possibilities for a negotiated peaceful settlement, without foreign interference.

In *Foreign Affairs*, Barnett Rubin and Ahmed Rashid recommend that U.S. strategy in the region should shift from more troops in Afghanistan and attacks in Pakistan to a "diplomatic grand bargain—forging compromise with insurgents while addressing an array of regional rivalries and insecurities."

The current military focus "and the attendant terrorism," they warn, might lead to the collapse of nuclear-armed Pakistan, with grim consequences. They urge the incoming U.S. administration "to put an end to the increasingly destructive dynamics of the Great Game in the region" through negotiations that recognize the interests of the concerned parties within Afghanistan as well as Pakistan and Iran, but also India, China and Russia, who "have reservations about a NATO base within their spheres of influence" and about the threats "posed by the United States and NATO" as well as by al-Qaida and the Taliban.

The incoming U.S. president, they write, must end "Washington's keenness for 'victory' as the solution to all problems, and the United States' reluctance to involve competitors, opponents, or enemies in diplomacy."

Early on, at any number of points in the danger zone, the Obama administration could act to break the ominous cycle of violence.

Nightmare in Gaza

JANUARY 14, 2009

At this writing, as Israel's vicious assault on Gaza continues, as the horror deepens and becomes ever bloodier, the prospects for a decent resolution fade amid the cries of the wounded, the dying and the grieving.

The latest assault on Gaza opened with Israel's violation of a cease-fire on November 4 [2008], as U.S. voters were going to the polls to elect Barack Obama, then broke out in full fury on December 27 [2008].

To these crimes Obama's response has been silence—unlike, say, the terrorist attacks in Mumbai, which he was quick to denounce, along with the "hateful ideology" that lies behind it. In the case of Gaza, his spokespersons hid behind the mantra that "there is one president at a time," and repeated his support for Israeli actions when he visited the Israeli town of Sderot in July: "If somebody was sending rockets into my house, where my two daughters sleep at night, I'm going to do everything in my power to stop that."

But he seemingly will do nothing, not even make a statement, when U.S. jets and helicopters with Israeli pilots are causing incomparably worse suffering to Palestinian children.

An incoming president's voice should be heard, at least as a moral response and a call to terminate crimes and aid humanitarian efforts, particularly in a case like this. Plainly,

Israel cannot act independently of its major ally, backer, arms supplier and enabler. The bloodshed in Gaza is on U.S. hands as well. That voice might have also indicated whether Washington's hard-line support for Israeli actions may soften, after decades.

During the campaign, rumors circulated that Obama might depart from the U.S. rejectionism that has long been a primary barrier to a genuine two-state settlement, an independent Palestinian state coexisting with Israel. This is the long-standing international consensus that the United States and Israel have blocked in practice, in virtual isolation, for more than thirty years, with rare and temporary departures. Otherwise the consensus is supported by the entire world, including the majority of the U.S. population. Obama's record, however, promises no basis for taking the rumors seriously.

As a special assistant on the Middle East, Obama has picked Daniel C. Kurtzer, Clinton-Bush ambassador to Egypt and Israel. Kurtzer took part in writing Obama's speech to the Israeli lobbying organization AIPAC in Washington last June [2008]. The remarkable text went well beyond President Bush in its obsequiousness, even declaring that "Jerusalem will remain the capital of Israel, and it must be undivided"—a position so extreme that his campaign had to explain that his words didn't mean what he said. And Kurtzer is one of the more moderate of Obama's choices on the region.

Israeli president Shimon Peres informed the press that on his July [2008] visit to Israel, Obama had said he was "very impressed" with the Arab League Proposal, which calls for full normalization of relations with Israel. That goes even beyond the consensus on a two-state settlement.

(Peres himself did not accept the consensus. Indeed, in his last days as prime minister in 1996, he held that a Palestinian state can never come into existence.)

Obama's reported comments might suggest a significant change of heart, except that on the same trip, as right-wing Israeli leader Benjamin Netanyahu reported, Obama had told him he was "very impressed" with Netanyahu's plan, which calls for indefinite Israeli control of the occupied territories.

The paradox is plausibly resolved by Israeli political analyst Aluf Benn, who points out that Obama's "main goal was not to screw up or ire anyone. Presumably he was polite, and told his hosts that their proposals were 'very interesting'—they leave satisfied and he hasn't promised a thing." Understandable, but it leaves us with nothing except Obama's fervent professions of love for Israel and disregard for Palestinian concerns.

Obama has long supported Israel's "Right to Self-Defense" and its "right to protect its citizens." In 2006, during Israel's U.S.-backed invasion of Lebanon, Obama co-sponsored "a senate resolution against Iran and Syria's involvement in the war, and insisting that Israel should not be pressured into a cease-fire that did not deal with the threat of Hezbollah missiles," according to his campaign website. The invasion of Lebanon, Israel's fifth, killed more than 1,000 Lebanese and once again destroyed much of southern Lebanon as well as parts of Beirut.

This happens to be the sole mention of Lebanon among foreign policy issues on the Obama website. Evidently, Lebanon has no right of self-defense. In fact, who could have a right of self-defense against the United States or its clients?

The invasion of Gaza is the latest tragic episode that follows from a peaceful democratic Palestinian election there in January 2006, carefully monitored and pronounced to be free and fair by international observers. But Palestinians voted for Hamas, despite U.S.-Israeli efforts on behalf of Palestinian Authority president Mahmoud Abbas and his Fatah party. Those who disobey the master must suffer for this misdeed.

The punishment of Palestinians for voting the wrong way began at once, and has grown increasingly more severe. With U.S. backing, Israel stepped up its violence in Gaza, kidnapped much of the elected leadership, steadily tightened its siege and even cut off the flow of water to the Gaza Strip. The United States and Israel made sure that the elected government would not have a chance to function.

Even when Israel formally accepted a cease-fire, as it did in June 2008, it instantly violated it, maintaining its siege (an act of war) and preventing UNRWA, the U.N. agency that keeps Palestinians alive, from replenishing its stores. "So when the cease-fire broke down [on November 4, 2008], we ran out of food for the 750,000 who depend on us," the director of UNRWA in Gaza, John Ging, informed the BBC.

In the weeks that followed, the blockade was tightened further, with disastrous consequences for the population. Both sides escalated violence (all deaths were Palestinian), until the cease-fire formally ended on December 19 [2008] and Prime Minister Ehud Olmert authorized the full-scale invasion, rejecting Hamas offers to extend it.

In the midst of yet another nightmare in Gaza, the hope for a two-state settlement in Israel-Palestine in accord with the international consensus, may seem almost unimaginable. Yet it is well to remember that it came rather close in January 2001, as already discussed, thanks to the willingness of a U.S. president to contemplate it.

Barack Obama and Israel-Palestine

FEBRUARY 3, 2009

Barack Obama is recognized to be a person of acute intelligence, a legal scholar, careful with his choice of words. He deserves to be taken seriously—both what he says, and what he omits.

Particularly significant is his first substantive statement on foreign affairs, on January 22 [2009], at the State Department, when introducing George Mitchell to serve as his special envoy for Middle East peace.

Mitchell is to focus his attention on the Israel-Palestine problem, in the wake of the recent U.S.-Israeli invasion of Gaza. During the murderous assault, Obama remained silent apart from a few platitudes, because, he said, there is only one president.

On January 22, however, the one president was Barack Obama, so he could speak freely about these matters. Obama emphasized his commitment to a peaceful settlement: "It will be the policy of my administration to actively and aggressively seek a lasting peace between Israel and the Palestinians, as well as Israel and its Arab neighbors."

But he left vague his policy's contours, apart from one specific proposal: "The Arab peace initiative," Obama said, "contains constructive elements that could help advance these efforts. Now is the time for Arab states to act on the initiative's promise by supporting the Palestinian government under President Abbas and Prime Minister Fayyad, taking steps toward normalizing relations with Israel, and by standing up to extremism that threatens us all."

Obama is not directly falsifying the Arab League proposal, but his carefully framed interpretation is instructive.

The proposal indeed calls for normalization of relations with Israel—in the context, it must be noted, and only in the context, of a two-state settlement, a prospect that the United States and Israel have blocked, virtually alone, for more than thirty years.

Obama's omission of that crucial fact—Israel and Palestine as coexisting states on the international border, with perhaps minor and mutual modifications—can hardly be accidental. It signals that he envisions no departure from U.S. rejectionism. His call for the Arab states to act on a corollary to their proposal, while the United States ignores even the existence of its central content, the precondition for the corollary, surpasses cynicism.

On the ground, the most significant acts that undermine a peaceful settlement are the daily U.S.-backed actions in the occupied territories, all recognized to be criminal: taking over valuable land and resources and constructing what the leading architect of the plan, Ariel Sharon, called "Bantustans" for Palestinians.

But the United States and Israel continue to oppose a political settlement even in words, most recently in December, when they (and a few Pacific islands) voted against a U.N. resolution supporting "the right of the Palestinian people to self-determination" (passed 173 to 5).

In referring to the "constructive" proposal, Obama had not one word to say about the settlement and infrastructure developments in the West Bank, and the complex measures to control Palestinian existence, designed to undermine the prospects for a peaceful two-state settlement. His silence refutes his oratorical flourishes about how "I will sustain an active commitment to seek two states living side by side in peace and security."

Obama persists in restricting support to Abbas and Fayyad, who represent the defeated parties in the free election of January 2006, which elicited instant and overt punishment by the United States and Israel. Obama's insistence that only Abbas and Fayyad exist conforms to the consistent Western contempt for democracy unless it is under control.

Obama also provided the usual reasons for ignoring the elected government led by Hamas. "To be a genuine party to peace," Obama declared, "the quartet [the United States, European Union, Russia and the United Nations] has made it clear that Hamas must meet clear conditions: recognize Israel's right to exist; renounce violence; and abide by past agreements."

Unmentioned, as usual, is the inconvenient fact that the United States and Israel bar a two-state settlement, virtually alone; they of course do not renounce violence; and they reject the quartet's central proposal, the "Road Map," as already discussed.

It is perhaps unfair to criticize Obama for this further exercise of cynicism, because it is close to universal.

Also near universal are the standard references to Hamas: a terrorist organization, dedicated to the destruction of Israel (or maybe all Jews). Omitted is that, unlike the two rejectionist states, Hamas has called for a two-state settlement in terms of the international consensus: publicly, repeatedly, explicitly.

Obama said, "Let me be clear: America is committed

to Israel's security. And we will always support Israel's right to defend itself against legitimate threats."

There was nothing about the right of Palestinians to defend themselves against far more extreme threats, such as those occurring daily, with U.S. support, in Gaza and the occupied territories. But that again is the norm.

The deceit is particularly striking in this case because of the occasion of Mitchell's appointment. Mitchell's primary achievement was his leading role in the peaceful settlement in northern Ireland. It called for an end to IRA terror and British violence, recognizing implicitly that while Britain had the right to defend itself from terror, it had no right to do so by force, because there was a peaceful alternative: recognition of the legitimate grievances of the Irish Catholic community that were the roots of IRA terror. When Britain adopted that sensible course, the terror ended.

Mitchell himself might welcome a serious two-state proposal. In 2001, for the George W. Bush administration, he chaired an international panel whose report at least barred any further Israeli settlement activity on the West Bank. The Mitchell Report, while formally accepted and praised by the United States and Israel, was completely ignored.

For Mitchell's new mission with regard to Israel-Palestine, the implications of Obama's remarks are obvious: A genuine two-state settlement isn't on the table.

Mitchell's first mandate for the Middle East is to open discussions, and to listen to everyone—everyone except for the elected government in Palestine. Obama's omissions are a striking indication of the commitment of his administration to traditional U.S. rejectionism and opposition to peace, except on its own extremist terms.

Latin America, Defiant

MARCH 8, 2009

More than a millennium ago, long before the European Conquest, a lost civilization flourished in the area now known as Bolivia.

Archaeologists are discovering that Bolivia had a wealthy, sophisticated and complex society—to quote their words, "one of the largest, strangest and most ecologically rich artificial environments on the face of the planet. . . . Their villages and towns were spacious and formal," creating a landscape that was "one of humankind's greatest works of art, a masterpiece."

Now Bolivia, along with much of the region from Venezuela to Argentina, is resurgent. The Conquest and its echo in U.S. imperial dominance in the hemisphere are giving way to independence and interdependence that mark a new dynamic in North-South relations, all against the backdrop of the meltdown in the U.S. and world economy.

During the past decade, Latin America has become the most progressive region in the world.

Initiatives throughout the subcontinent have had a significant impact in individual countries and in the slow emergence of regional institutions.

Among these are the Banco del Sur, endorsed in 2007 by Nobel laureate economist Joseph Stiglitz in Caracas, Venezuela; and the ALBA, the Bolivarian Alternative for

Latin America and the Caribbean, which might prove to be a true dawn if its initial promise can be realized.

The ALBA is often described as an alternative to the U.S.-sponsored "Free Trade Area of the Americas," but the terms are misleading. It should be understood as an independent development, not an alternative. And, furthermore, the so-called "free trade agreements" have only a limited relation to free trade, or even to trade in any serious sense of that term; and they are certainly not agreements, at least if people are part of their countries.

A more accurate term would be "investor-rights arrangements," designed by multinational corporations and banks and the powerful states that cater to their interests, established mostly in secret, without public participation or awareness.

Another promising regional organization is UN-ASUR, the Union of South American Nations. Modeled on the European Union, UNASUR aims to establish a South American parliament in Cochabamba, Bolivia, a fitting site: In 2000, the people of Cochabamba staged a courageous and successful struggle against privatization of water—which awakened international solidarity as a proof of what can be achieved by committed activism.

The southern-cone dynamic has in part flowed from Venezuela, with the election of Hugo Chávez, a leftist president dedicated to using Venezuela's rich resources for the benefit of the Venezuelan people rather than for wealth and privilege at home and abroad, and to promote the regional integration so desperately needed as a prerequisite for independence, for democracy and for meaningful development.

Chávez is far from alone in these goals. Bolivia, the

poorest country in the continent, is perhaps the most dramatic example.

Bolivia has blazed an important path to true democratization in the hemisphere. In 2005, the indigenous majority, the most repressed population in the hemisphere, entered the political arena and elected someone from their own ranks, Evo Morales, to pursue programs that derived from popular organizations.

The election was only one stage in ongoing struggles. The issues were well known and serious: control over resources, cultural rights and justice in a complex multiethnic society, and the huge economic and social gap between the great majority and the narrow wealthy elite, the traditional rulers.

In consequence, Bolivia also is the scene of today's most dangerous confrontation between popular democracy and privileged Europeanized elites who resent the loss of their political privilege and thus oppose democracy and social justice, sometimes violently. Routinely, they enjoy firm U.S. backing.

Last September [2008], at an emergency UNASUR summit in Santiago, Chile, South American leaders declared "their full and firm support for the constitutional government of President Evo Morales, whose mandate was ratified by a big majority"—referring to his victory in the recent referendum.

Morales thanked UNASUR, observing that "for the first time in South America's history, the countries of our region are deciding how to resolve our problems, without the presence of the United States."

The United States had long dominated Bolivia's economy, especially for processing its tin exports. As international affairs scholar Stephen Zunes points out, in the early 1950s, "at a critical point in the nation's effort to become more self-sufficient, the U.S. government forced Bolivia to use its scarce capital not for its own development, but to compensate the former mine owners and repay its foreign debts."

The economic policies forced on Bolivia at that time were a precursor of the structural-adjustment programs imposed on the continent thirty years later, under the terms of the neoliberal "Washington consensus," which has generally had disastrous effects wherever its strictures have been observed.

By now, the victims of neoliberal market fundamentalism are coming to include the rich countries, where the curse of financial liberalization has helped to bring about the worst financial crisis since the Great Depression.

The traditional modalities of imperial control—violence and economic war—are weakening. Latin America has real choices. Washington well understands that these choices threaten not only its domination of the hemisphere, but also its global dominance. Control of Latin America has been a goal of U.S. foreign policy since the first days of the Republic.

If the United States could not control Latin America, it could not expect "to achieve a successful order elsewhere in the world," Nixon's National Security Council concluded in 1971 while considering the paramount importance of destroying Chilean democracy, as it did.

Mainstream scholarship recognizes that Washington has supported democracy if and only if it contributes to strategic and economic interests, a policy that continues without change through all administrations, to the present.

These antidemocratic concerns are the rational form of the domino theory, sometimes more accurately called "the threat of a good example." For such reasons, even the tiniest departure from strict obedience is regarded as an existential threat that calls for a harsh response: peasant organizing in remote communities of northern Laos, fishing cooperatives in Grenada and so on throughout the world.

In a newly self-confident Latin America, integration has at least three dimensions—regional, a crucial prerequisite to independence, making it more difficult for the master of the hemisphere to pick off countries one by one; global, in establishing South-South relations and diversifying markets and investment, with China an increasingly significant partner in hemispheric affairs; and internal, perhaps the most vital dimension of all. Latin America is notorious for its extreme concentrations of wealth and power, and privileged elites' lack of responsibility for the welfare of the nation.

Latin America has huge problems, but there are many promising developments that may herald an era of true globalization—international integration in the interests of people, not investors and other concentrations of power.

Down with the Durand Line!

APRIL 1, 2009

Since antiquity, the region now known as Afghanistan has been a crossroads for would-be conquerors. Alexander the Great, Genghis Khan and Tamerlane reigned there.

During the nineteenth century, the British and the Russian empires jockeyed for supremacy in Central Asia—the Great Game, as the rivalry was described. In 1893, Sir Henry Mortimer Durand, a British colonial officer, drew a 1,500-mile line to define the western edge of British-ruled India. The Durand Line cut through Pashtun tribal areas that Afghans considered part of their country. In 1947, the northwest part of the region was carved into the new state of Pakistan.

The Great Game continues in Afghanistan-Pakistan—Afpak, as it's now called. The term makes sense for the region on either side of the faint, porous Durand Line, which the population never accepted and the state of Afghanistan, when it was still functioning, consistently opposed.

One indelible historical marker is that the Afghans have vigorously fought off all invaders.

Afghanistan continues to be a geostrategic prize in the Great Game. In Afpak, President Obama has proceeded, in accord with his campaign promises, to step up the war considerably, carrying forward the Bush administration's pattern of escalation.

Currently, Afghanistan is occupied by the United States

and its NATO allies. The outsiders' military presence only arouses confrontations, whereas what is needed is a common effort among concerned regional powers—including China, India, Iran, Pakistan and Russia—that would help Afghans face their internal problems peacefully, as many believe they can.

NATO has moved far beyond its Cold War origins. After the Soviet Union collapsed, NATO lost its pretext for existence: defense against a hypothetical Russian assault. But NATO quickly took on new missions, expanding to the east in violation of promises to Mikhail Gorbachev, a serious security threat to Russia, naturally raising international tensions.

President Obama's national security adviser, James Jones, NATO supreme allied commander in Europe from 2003 to 2006, advocates NATO expansion east and south, steps that would reinforce U.S. control over Middle East energy supplies (in technical terms, "safeguarding energy security"). He also champions a NATO response force, which will give the U.S.-run military alliance "much more flexible capability to do things rapidly at very long distances."

China may represent Washington's greatest concern. The China-based Shanghai Cooperation Organization, which some analysts regard as a potential counterbalance to NATO, includes Russia and the Central Asian states. India, Pakistan and Iran are observers, and there is speculation about their joining. China has also deepened relations with Saudi Arabia, the jewel in the crown of the oil system.

A grassroots counterforce to the great-power maneuvering is the strong peace movement that is growing in Afghanistan. Activists have called for an end to violence and negotiations with the Taliban. These Afghans welcome outside help—for reconstruction and development, not military purposes.

The peace movement is gathering so much popular support in Afghanistan that the U.S. troops pouring into the country will face not only the Taliban but also "an unarmed but equally daunting foe: public opinion," reports Pamela Constable of the *Washington Post* on her recent visit to Afghanistan. Many Afghans say that "instead of helping to defeat the insurgents and quell the violence that has engulfed their country, more foreign troops will exacerbate the problem."

Most of the Afghans interviewed by Constable "said they would prefer a negotiated settlement with the insurgents to an intensified military campaign. Several pointed out that the Taliban fighters are fellow Afghans and Muslims, and that the country has traditionally settled conflicts through community and tribal meetings."

Afghan President Hamid Karzai's first message to Obama, apparently unanswered, was a request to stop attacking civilians. Karzai also informed a U.N. delegation that he wants a timetable for withdrawal of foreign (meaning U.S.) troops. He is therefore out of favor in Washington, and has accordingly shifted from a media favorite to "unreliable," "corrupt," etc.—no more true than when he was feted as our "our man" in Kabul. The press reports that the United States and its allies are planning to sideline him in favor of a figure of their choice. Karzai's popularity has also declined in Afghanistan, though it remains far above that of the American occupying forces.

A useful perspective comes from the experienced British correspondent Jason Burke, who writes, "we are still hoping to build the state we want the Afghans to want, rather than the state that they actually want. Ask many Afghans which state they hope their own will resemble in a few decades and the answer is 'Iran.'"

Iran's role is particularly important. It has intimate relations with Afghanistan. It strongly opposes the Taliban and gave substantial help in expelling them—to be rewarded by being branded part of the Axis of Evil. It has more interest in a stable and flourishing Afghanistan than any other country, and has natural relations with Pakistan, India, Turkey, China and Russia. These relations may well develop on their own, perhaps associated with the Shanghai Cooperation Organization, if the United States continues to block Iran's relations with the Western world.

This week [March 2009], at a U.N. conference on Afghanistan in The Hague, Karzai met Iranian officials who pledged to help with reconstruction and to cooperate on regional policing of the booming Afghan drug trade.

The Bush-Obama policy of escalation is not conducive to a peaceful settlement in Afghanistan or the region. What is important is negotiations among Afghans without foreign interference, Great Game or otherwise. The problems of Afghanistan are matters for Afghans to settle.

A Tradition of Torture

MAY 5, 2009

The torture memos released by the White House have elicited shock, indignation and surprise. The shock and indignation are understandable—particularly the just-declassified Senate Armed Services Committee Report on Detainee Treatment.

In the summer of 2002, as the report reveals, interrogators at Guantánamo came under increasing pressure from up the chain of command to establish a link between Iraq and al-Qaida. Waterboarding, among other measures of torture, finally elicited the "evidence" from a detainee that was used to help justify the Bush-Cheney invasion of Iraq the next year.

But why the surprise about the torture memos? Even without inquiry, it was reasonable to suppose that Guantánamo was a torture chamber. Why else send prisoners where they would be beyond the reach of the law—incidentally, a place that Washington is using in violation of a treaty that was forced on Cuba at the point of a gun? The security rationale is hard to take seriously.

A broader reason why there should be little surprise is that torture has been a routine practice from the early days of the conquest of the national territory, and then beyond, as the imperial ventures extended to the Philippines, Haiti and elsewhere.

Furthermore, torture was the least of the many crimes of aggression, terror, subversion and economic strangulation that have darkened U.S. history, much as they have done for other great powers. The current torture revelations yet again point up the perennial conflict between "what we stand for" and "what we do."

The reaction has been vehement but in ways that raise some questions. For example, *New York Times* columnist Paul Krugman, one of the most eloquent and forthright critics of Bush malfeasance, writes that we used to be "a nation of moral ideals," and never before Bush "have our leaders so utterly betrayed everything our nation stands for."

To say the least, that common view is a rather slanted version of history. It is an article of faith, almost a part of the national creed, that the United States is righteously unlike other great powers, past and present—the notion that is called "American exceptionalism."

A partial corrective might be British journalist Godfrey Hodgson's just-published history, *The Myth of American Exceptionalism*. Hodgson concludes that the United States is "just one great, but imperfect, country among others."

International Herald Tribune columnist Roger Cohen, reviewing the book in the *New York Times*, agrees that the evidence supports Hodgson's judgment but takes issue with him on a fundamental point: Hodgson fails to understand that "America was born as an idea, and so it has to carry that idea forward."

The idea is revealed by America's birth as a "city on a hill," Cohen writes, an "inspirational notion" that resides "deep in the American psyche."

In brief, Hodgson's error is that he is keeping to "the distortions of the American idea in recent decades" (in fact, throughout its history). Let us turn then to the "idea" of America.

The inspirational phrase "city on a hill" was coined by John Winthrop in 1630, borrowing from the Gospels and outlining the glorious future of a new nation "ordained by God."

One year earlier his Massachusetts Bay Colony established its Great Seal. It depicts an Indian with a scroll coming out of his mouth. On it are the words, "Come over and help us." The British colonists were thus benevolent humanists, responding to the pleas of the miserable natives to be rescued from their bitter pagan fate.

This early proclamation of "humanitarian intervention," to use the currently popular term, turned out very much like its successors, though unusually horrifying for the beneficiaries, as there should be no need to review.

There are sometimes innovations. During the past sixty years, victims worldwide have endured what historian Alfred McCoy describes as the CIA's "revolution in the cruel science of pain" in his 2006 book, *A Question of Torture: CIA Interrogation, from the Cold War to the War on Terror.* As he discusses, often the task of torture is farmed out to subsidiaries, unlike the current version. And waterboarding is one of the decades-old methods that appear with little change at Guantánamo.

Complicity in torture often features in U.S. foreign policy. In a 1980 study, political scientist Lars Schoultz found that U.S. aid "has tended to flow disproportionately to Latin American governments which torture their citizens . . . to the hemisphere's relatively egregious violators of fundamental human rights."

The Schoultz study and others reaching similar conclusions precede the Reagan years, when the topic was not worth studying because the correlations were so overwhelmingly clear. And that tendency continues to the present without significant modification.

Small wonder that the president advises us to look forward, not backward—a convenient doctrine for those who hold the clubs. Those who are beaten by them tend to see the world differently, much to our annoyance.

Among empires, "exceptionalism" is probably close to universal. France was hailing its "civilizing mission" while the French Minister of War called for "exterminating the indigenous population" of Algeria.

Britain's nobility was a "novelty in the world," John Stuart Mill declared, while urging that this angelic power delay no longer in completing its liberation of India. Mill's classic essay, "A Few Words about Non-Intervention," was written right after the public revelation of Britain's horrifying atrocities in suppressing the 1857 Indian rebellion.

Such "exceptionalist" ideas are not only convenient for power and privilege, but also pernicious. One reason is that they efface real ongoing crimes. The My Lai massacre during the Vietnam War was a mere footnote to the vastly greater atrocities of the post-Tet pacification programs. The Watergate break-in that brought down a U.S. president was doubtless criminal, but the furor over it displaced incomparably worse crimes at home and abroad—the bombing of Cambodia, to mention just one horrific example. Quite commonly, selectively chosen atrocities have this function.

Historical amnesia is a very dangerous phenomenon, not only because it undermines moral and intellectual integrity, but also because it lays the groundwork for crimes that lie ahead.

Barack Obama and Israel-Palestine

JUNE 4, 2009

A CNN headline on Obama's plans for his Cairo address reads, "Obama looks to reach the soul of the Muslim world." Perhaps that captures his intent, but more significant is the content that the rhetorical stance hides—or more accurately, omits.

On Israel-Palestine—the speech offered nothing substantive. Obama called on Arabs and Israelis not to "point fingers" at each other or to "see this conflict only from one side or the other."

Crucially omitted is a third side, the United States, which has played a decisive role in sustaining the conflict. Obama did not indicate that the U.S. role should change or even be considered.

Obama once again praised the Arab Peace Initiative, saying that Arabs should see it as "an important beginning, but not the end of their responsibilities." Again there is a crucial omission, surely conscious: Obama and his advisers are undoubtedly aware that the Initiative reiterates the long-standing international consensus for a two-state settlement on the international (pre-June 1967) border, perhaps with "minor and mutual modifications," to borrow the U.S. government language before the United States had departed from the world on this issue thirty-five years ago. It is in the context of the international consensus that the Arab Peace Initiative calls on Arab states to normalize relations with Israel.

Obama asked the Arab states to proceed with normalization, studiously ignoring the crucial political settlement that is its precondition. The Initiative cannot be a "beginning" if the United States continues to refuse to accept its core principles, even to acknowledge them.

What is Israel to do in return for the Arab states' taking steps to normalize relations? The strongest position so far enunciated by the Obama administration is that Israel should conform to Phase I of the 2003 Road Map: "Israel freezes all settlement activity (including natural growth of settlements)."

Overlooked in the debate about the settlements is that even if Israel were to accept Phase I, that would leave in place the entire settlement project that has already been developed, with key U.S. support. The settlements and development projects ensure that Israel will take over the valuable land within the illegal "separation wall" (including major water supplies of the region) as well as the Jordan Valley, thus imprisoning Palestinians within a limited territory that is, furthermore, being broken up into cantons by settlement/infrastructure salients extending far to the east.

Unmentioned as well is that Israel is taking over Greater Jerusalem, the site of its major current development programs, displacing many Arabs, so that what remains to Palestinians will be separated from the center of their cultural, economic and sociopolitical life.

All this settlement activity is in violation of international law, particularly so in the case of Jerusalem, since it is in violation of early and explicit Security Council resolutions.

The Bush I administration went a bit beyond words in objecting to illegal Israeli settlement projects, namely, by withholding U.S. economic support for them. In contrast, Obama administration officials stated that such measures are "not under discussion" and that any pressures on Israel to conform to the Road Map will be "largely symbolic," the *New York Times* reported.

In the background of the Middle East trip is the Obama administration's goal, enunciated most clearly by Senator John Kerry, chair of the Senate Foreign Relations Committee, to forge an alliance of Israel and the "moderate" Arab states against Iran. Such an alliance would serve as a bulwark for U.S. domination of the vital energy-producing regions.

(The term "moderate," as usual, has nothing to do with the character of the state, but rather signals its willingness to conform to U.S. demands.)

The unparalleled services that Israel provides for the U.S. military and intelligence agencies, and for high-tech industry, afford it a certain leeway to defy Washington's orders—though it is taking a chance of offending its patron. The extremism of the current Israeli government has been held in check by more sober elements.

If Israel ever goes too far, there might indeed erupt a U.S.-Israel policy showdown of the kind that many commentators perceive today—so far, with little factual basis, in the Cairo speech or elsewhere. The forecast for U.S. policy in Israel-Palestine is likely to be more of the same.

A Season of Travesties

The elections in Lebanon and Iran and the coup in Honduras are significant not only in themselves but also in the international reactions to them. The comparative non-reaction to a current act of Israeli piracy in the Mediterranean is a footnote.

LEBANON

The June 7 [2009] election was greeted with euphoria in the mainstream.

"I'm a sucker for free and fair elections," *New York Times* columnist Thomas Friedman wrote on June 10 [2009]. "In Lebanon it was real deal, and the results were fascinating: President Barack Obama defeated President Mahmoud Ahmadinejad of Iran."

The proof is that "a solid majority of all Lebanese—Muslims, Christians and Druse—voted for the March 14 coalition led by Saad Hariri," the U.S.-backed candidate and son of the murdered ex–Prime Minister Rafik Hariri.

We must give credit where it is due for this triumph of free elections (and of Washington), Friedman continues: "Without George Bush standing up to the Syrians in 2005—and forcing them to get out of Lebanon after the Hariri killing—this free election would not have happened. Bush created the space. [In his Cairo speech] Obama helped stir the hope."

Two days later, Friedman's praise for our noble role in the Middle East was echoed in a *Times* op-ed by Elliott Abrams, a senior fellow at the Council on Foreign Relations, formerly a high official under President Reagan and Bush II: "The voting in Lebanon passed any realistic test. . . . The Lebanese had a chance to vote against Hezbollah, and took the opportunity."

A "realistic test," however, might include the actual vote. The Hezbollah-based March 8 coalition won handily, by approximately the same figure as Obama vs. McCain in November [2008], about 54 percent of the popular vote, according to Lebanese Ministry of Interior figures.

Hence by the Friedman-Abrams argument, we should be lamenting Ahmadinejad's defeat of Obama—at least those who have minimal respect for democracy.

Like others, Friedman and Abrams are referring to representatives in Parliament. These numbers are skewed by Lebanon's confessional voting system, which sharply reduces the number of seats granted to the largest of the sects, the Shiites, who overwhelmingly back Hezbollah and its ally Amal.

As serious analysts have pointed out, Lebanon's "confessional" ground rules also undermine "free and fair elections" in even more significant ways. Political analyst Assaf Kfoury observes that the ground rules leave no space for non-sectarian parties and erect a barrier to introducing socioeconomic policies and other real issues into the electoral system.

Accordingly these rules open the door to "massive external interference," low voter turnout and "vote-rigging and vote-buying," all features of the June election, even more so than before.

Thus in Beirut, home of nearly half of Lebanon's population, less than a fourth of eligible voters could vote without returning to their usually remote districts of origin. The effect is that migrant workers and the poorer classes are effectively disenfranchised in "a form of extreme gerrymandering, Lebanese-style," favoring the privileged and pro-Western classes.

IRAN

As in Lebanon, Iran's electoral system itself violates basic rights. Candidates have to be approved by ruling clerics, who can and do bar advocates of policies of which they disapprove.

The electoral results from Iran's Interior Ministry lacked credibility both by the manner of their release and by the figures themselves—touching off an enormous popular protest brutally suppressed by the armed forces of the ruling clerics. Perhaps Ahmadinejad might have won a majority if the votes were fairly counted, but the rulers apparently weren't willing to take the chance.

From the streets of Tehran, correspondent Reese Erlich writes, "It's a genuine Iranian mass movement made up of students, workers, women and middle-class folks"—and possibly much of the rural population.

Eric Hooglund, a scholar and expert on rural Iran, describes "overwhelming" support for opposition candidate Mir Hossein Mousavi among the people in the regions he has studied, and "palpable moral outrage over what came to be believed as the theft of their election."

It is highly unlikely that the protest will damage the clerical-military regime in the short term, but as Erlich observes, it "is sowing the seeds for future struggles."

ISRAEL-PALESTINE

We shouldn't forget one authentically "free and fair" recent
election in the Middle East—in Palestine in January 2006,
to which the United States and its allies at once responded
by punishing the population that voted "the wrong way."

Israel imposed a siege on Gaza and, last winter, merci-
lessly attacked it.

Relying on the impunity it receives as a U.S. client,
Israel has again enforced its blockade by hijacking the Free
Gaza movement boat *Spirit of Humanity* in international
waters and forcing it to the Israeli port of Ashdod.

The boat had left from Cyprus, where the cargo was
inspected: medicines, reconstruction supplies and toys.
The human rights workers aboard included Nobel Laureate Mairead Maguire and former congresswoman Cynthia
McKinney.

The crime scarcely elicited a yawn—with some justice,
one might argue, since Israel has been hijacking boats trav-
eling between Cyprus and Lebanon for decades. So why
even bother to report this latest outrage by a rogue state
and its patron?

HONDURAS

Central America is also the scene of an election-related
crime. A military coup in Honduras ousted President
Manuel Zelaya and expelled him to Costa Rica.

The coup replays what Latin American affairs analyst
Mark Weisbrot calls "a recurrent story in Latin America,"
pitting "a reform president who is supported by labor unions
and social organizations against a mafia-like, drug-ridden,
corrupt political elite accustomed to choosing not only the
Supreme Court and the Congress, but also the president."

Mainstream commentary describes the coup as an unfortunate return to the bad days of decades ago. But that is mistaken. This is the third military coup in the past decade, all conforming to the "recurrent story."

The first, in Venezuela in 2002, was supported by the Bush administration, which, however, backed down after sharp Latin American condemnation and restoration of the elected government by a popular uprising.

The second, in Haiti in 2004, was carried out by Haiti's traditional torturers, France and the United States. The elected president, Jean-Bertrand Aristide, was spirited to Central Africa.

What is novel in the Honduras coup is that Washington did not at once lend its overt support. Rather, the United States joined with the Organization of American States in opposing the takeover, though with a more mild condemnation than others, and with no action. Unlike the neighboring states, as well as France, Spain and Italy, the United States has not withdrawn its ambassador.

It surpasses the imagination that Washington didn't have advance knowledge of what was under way in Honduras, which is highly dependent on U.S. aid and whose military is U.S.-armed, -trained and -advised. Military relations have been close since the 1980s, when Honduras was the base for President Reagan's terrorist war against Nicaragua. Whether the "recurrent story" plays out again depends in no small measure on reactions within the United States.

(Regrettably, Obama did keep to the recurrent story, breaking from most of Latin America and Europe by recognizing an election organized by the coup regime in an atmosphere of repression and state violence.)

Making War to Bring "Peace"

JULY 29, 2009

A debate is under way at the United Nations over a policy that may seem uncontroversial: an international framework to prevent severe crimes against humanity.

The framework is called "responsibility to protect," or R2P, in common parlance. A version of R2P, adopted at the U.N. World Summit in 2005, reaffirmed rights and responsibilities that were accepted by member states in the past and sometimes implemented by them, changing little more than focus.

The discussions about R2P or its cousin, "humanitarian intervention," are regularly disturbed by the rattling of a skeleton in the closet: history, to the present.

Throughout history, few principles of international affairs apply generally. One is the maxim of Thucydides that the strong do as they wish while the weak suffer as they must.

Another principle is that virtually every use of force in international affairs has been accompanied by lofty rhetoric about the solemn responsibility to protect the suffering populations.

Understandably, the powerful prefer to forget history and look forward. For the weak, it is not a wise choice.

The skeleton in the closet made an appearance in the first dispute considered by the International Court of Justice (ICJ) sixty years ago, the Corfu Channel case, about an incident involving Great Britain and Albania.

The court determined it "can only regard the alleged right of intervention as the manifestation of a policy of force, such as has, in the past, given rise to most serious abuses and such as cannot, whatever be the defects in international organization, find a place in international law . . . ; from the nature of things, [intervention] would be reserved for the most powerful states, and might easily lead to perverting the administration of justice itself."

The same perspective informed the first meeting of the South Summit of 133 states in 2000. Its declaration, surely with the bombing of Serbia in mind, rejected "the so-called 'right' of humanitarian intervention, which has no legal basis in the United Nations Charter or in the general principles of international law."

The wording reaffirms the U.N. Declaration on Friendly Relations (1970). It has been repeated since, among others by the Ministerial Meeting of the Non-aligned movement in Malaysia in 2006, again representing the traditional victims in Asia, Africa, Latin America and the Arab world.

The same conclusion was drawn in 2004 by the high-level U.N. Panel on Threats, Challenges and Change, including Brent Scowcroft and other prominent Western diplomats. The panel concluded that U.N. Charter Article 51—which allows the use of force in self-defense under sharp restrictions, but in no other circumstances—"needs neither extension nor restriction of its long-understood scope."

The panel added, "For those impatient with such a response, the answer must be that, in a world full of perceived potential threats, the risk to the global order and the norm of nonintervention on which it continues to be based is simply too great for the legality of unilateral preventive action, as distinct from collectively endorsed action, to be accepted. Allowing one to so act is to allow all"—which is, of course, unthinkable.

The same basic position was adopted by the U.N. World Summit in 2005. The Summit also affirmed the willingness "to take collective action . . . through the Security Council, in accordance with the Charter . . . should peaceful means be inadequate and national authorities are manifestly failing to protect their populations" from serious crimes.

At most, the phrase sharpens the wording of Article 42 on authorizing the Security Council to resort to force. And the phrase keeps the skeleton in the closet—even if we were to regard the Security Council as a neutral arbiter, not subject to the maxim of Thucydides, an assumption that is untenable.

The Council is controlled by its five permanent members, and they are not equal in operative authority. One indication is the record of vetoes—the most extreme form of violation of a Security Council Resolution.

During the past quarter-century, China and France together vetoed seven resolutions; Russia, six; the United Kingdom, ten: and the United States, forty-five, including even resolutions calling on states to observe international law.

One way to mitigate this defect in the World Summit consensus would be to eliminate the veto, in accord with the will of the majority of the U.S. population. But such heresies are unthinkable, as much so as applying R2P right now to those who desperately need protection but are not on the favored list of the powerful.

There have been departures from the Corfu Channel restriction and its descendants. The Constitutive Act of the African Union asserts "the right of the Union to intervene in a Member State . . . in respect of grave circumstances." That differs from the Charter of the Organization of American States, which bars intervention "for any reason whatever, in the internal or external affairs of any other state."

The reason for the difference is clear. The OAS Charter seeks to deter intervention by the United States, but after the disappearance of the apartheid states, the African Union faces no similar problem.

I know of only one high-level proposal to extend R2P beyond the summit consensus and the African Union extension: the Report of the International Commission on Intervention and State Sovereignty on Responsibility to Protect (2001).

The Commission considers the situation in which "the Security Council rejects a proposal or fails to deal with it in a reasonable time." In that case, the report authorizes "action within area of jurisdiction by regional or sub-regional organizations . . . subject to their seeking subsequent authorization from the Security Council."

At this point, the skeleton in the closet rattles loudly. The powerful unilaterally determine their own "area of jurisdiction." The OAS and African Union cannot do so, but NATO can, and does, alone among organizations.

NATO has determined that its "area of jurisdiction" extends to the Balkans, Afghanistan and beyond.

The expansive rights accorded by the Report (which has no standing) are in practice restricted to NATO alone, violating the Corfu Channel principles and opening the door for R2P as a weapon of imperial intervention at will.

The "responsibility to protect" has always been selective. Thus it did not apply to the sanctions against Iraq imposed by the United States and United Kingdom and administered by the Security Council, condemned as "genocidal" by the distinguished diplomats in charge, both of whom resigned in protest.

There is also no thought today of applying R2P to the people of Gaza, a "protected population" for whom the United Nations is responsible. More generally, depredations by the U.S. and its allies and clients are immune.

And nothing serious is contemplated about the worst catastrophe in Africa, if not the world: the murderous conflict in eastern Congo. There, the BBC just reported, multinationals are once again accused of violating a U.N. resolution against illicit trade of valuable minerals—funding the violence.

Nor is R2P invoked to respond to massive starvation in the poor countries.

Several years ago UNICEF reported that sixteen thousand children die every day from lack of food, many more from easily preventable disease. The figures are higher now. In southern Africa alone it is Rwanda-level killing, not for a hundred days, but every day. Action under R2P would be easy enough, were there the will.

In these and numerous other cases the selectivity conforms to the maxim of Thucydides and the expectations of the ICJ sixty years ago.

But the maxims that largely guide international affairs are not immutable, and, in fact, have become less harsh over the years as a result of the civilizing effect of popular movements.

For such progressive reform, R2P can be a valuable tool, much as the Universal Declaration of Human Rights has been.

Even though states do not adhere to the Universal Declaration, and some formally reject much of it (crucially including the world's most powerful state), nonetheless it serves as an ideal that activists can appeal to in educational and organizing efforts, often effectively.

The discussion of R2P may be similar. With sufficient commitment, unfortunately not yet detectable among the powerful, it could be significant indeed.

Militarizing Latin America

SEPTEMBER 20, 2009

The United States was founded as an "infant empire," in the words of George Washington. The conquest of the national territory was a grand imperial venture. From the earliest days, control over the hemisphere was a critical goal.

Latin America has retained its primacy in U.S. global planning. If the United States cannot control Latin America, it cannot expect "to achieve a successful order elsewhere in the world," observed President Richard M. Nixon's National Security Council in 1971 when Washington was considering the overthrow of Salvador Allende's government in Chile.

Recently the hemisphere problem has intensified. South America has moved toward integration, a prerequisite for independence; has broadened international ties; and has begun to address persistent internal violations of elementary human rights.

The problem came to a head a year ago in Bolivia, South America's poorest country, where the new South American organization UNASUR supported the elected president Evo Morales against violent opposition by U.S.-backed traditional elites, as already discussed.

In another manifestation, Ecuador's president Rafael Correa vowed to terminate Washington's use of the Manta military base, the last such base open to the United States in South America, and later did.

In July [2009], the U.S. and Colombia concluded a secret deal to permit the United States to use seven military bases in Colombia.

The official purpose is to counter narcotics trafficking and terrorism, "but senior Colombian military and civilian officials familiar with negotiations" told the Associated Press "that the idea is to make Colombia a regional hub for Pentagon operations."

The agreement provides Colombia with privileged access to U.S. military supplies, according to reports. Colombia had already become the leading recipient of U.S. military aid (apart from Israel-Egypt, a separate category).

Colombia has had by far the worst human rights record in the hemisphere since the Central American wars of the 1980s. The correlation between U.S. aid and human rights violations has long been noted by scholarship.

The Associated Press also cited an April 2009 document of the U.S. Air Mobility Command, which proposes that the Palanquero base in Colombia could become a "cooperative security location."

From Palanquero, "nearly half the continent can be covered by a C-17 (military transport) without refueling," the document states. This could form part of "a global en route strategy," which "helps achieve the regional engagement strategy and assists with the mobility routing to Africa."

On August 28 [2009], UNASUR met in Bariloche, Argentina, to consider the U.S. military bases in Colombia.

After intense debate, the final declaration stressed that South America must be kept as "a land of peace," and that foreign military forces must not threaten the sovereignty or integrity of any nation of the region. And it instructed the South American Defense Council to investigate the Air Mobility Command document.

The bases' official purpose did not escape criticism. Morales said he witnessed U.S. soldiers accompanying Bolivian troops who fired at members of his coca growers union.

"So now we're narco-terrorists," he continued. "When they couldn't call us communists anymore, they called us subversives, and then traffickers, and since the September 11 attacks, terrorists." He warned that "the history of Latin America repeats itself."

The ultimate responsibility for Latin America's violence lies with U.S. consumers of illegal drugs, Morales said: "If UNASUR sent troops to the United States to control consumption, would they accept it? Impossible."

Last February [2009], the Latin American Commission on Drugs and Democracy issued its analysis of the U.S. "war on drugs" in past decades.

The commission, led by former Latin American presidents Fernando Cardoso (Brazil), Ernesto Zedillo (Mexico), and César Gaviria (Colombia), concluded that the drug war had been a complete failure and urged a drastic change of policy, away from forceful measures at home and abroad and toward much less costly and more effective measures—prevention and treatment.

The commission report, like earlier studies and the historical record, had no detectable impact. The nonresponse reinforces the natural conclusion that the "drug war"—like the "war on crime" and "the war on terror"—is pursued for reasons other than the announced goals, reasons that are revealed by the consequences.

Under the pretexts of the drug war, during the past decade, the United States has increased military aid and training of Latin American officers in light infantry tactics to combat "radical populism"—a concept that, in the Latin American context, sends shivers up the spine.

Military training is being shifted from the State Department to the Pentagon, eliminating human rights and democracy provisions formerly under congressional supervision, always weak but at least a deterrent to some of the worst abuses.

The U.S. Fourth Fleet, disbanded in 1950, was reactivated in 2008, shortly after Colombia's invasion of Ecuador, with responsibility for the Caribbean, Central and South America, and the surrounding waters.

Its "various operations . . . include counter-illicit trafficking, Theater Security Cooperation, military-to-military interaction and bilateral and multinational training," the official announcement says.

Militarization of South America aligns with much broader designs. In Iraq, information is virtually nil about the fate of the huge U.S. military bases there, so they presumably remain for force projection. The cost of the immense city-with-in-a-city embassy in Baghdad is to rise to $1.8 billion a year, from an estimated $1.5 billion.

The Obama administration is also building mega-embassies in Pakistan and Afghanistan.

The United States and United Kingdom are demanding that the U.S. military base in Diego Garcia be exempted from the planned African nuclear-weapons-free zone—as U.S. bases are off-limits in similar zoning efforts in the Pacific.

In short, moves toward "a world of peace" do not fall within the "change you can believe in," to borrow Obama's campaign slogan.

War, Peace and Obama's Nobel

OCTOBER 26, 2009

The hopes and prospects for peace aren't well aligned—not even close. The task is to bring them nearer. Presumably that was the intent of the Nobel Peace Prize committee in choosing President Barack Obama.

The prize "seemed a kind of prayer and encouragement by the Nobel committee for future endeavor and more consensual American leadership," Steven Erlanger and Sheryl Gay Stolberg wrote in the *New York Times*.

The nature of the Bush-Obama transition bears directly on the likelihood that the prayers and encouragement might lead to progress.

The Nobel committee's concerns were valid. They singled out Obama's rhetoric on reducing nuclear weapons.

Right now Iran's nuclear ambitions dominate the headlines. The warnings are that Iran may be concealing something from the International Atomic Energy Agency (IAEA) and violating U.N. Security Council Resolution 1887, passed last month [September 2009], hailed as a victory for Obama's efforts to contain Iran.

Meanwhile, a debate continues on whether Obama's recent decision to reconfigure missile-defense systems in Europe is a capitulation to the Russians or a pragmatic step to defend the West from Iranian nuclear attack.

Silence is often more eloquent than loud clamor, so let us attend to what is unspoken.

Amid the furor over Iranian duplicity, the IAEA passed a resolution calling on Israel to join the Nuclear Non-Proliferation Treaty (NPT) and open its nuclear facilities to inspection.

The United States and Europe tried to block the IAEA resolution, but it passed anyway. The media virtually ignored the event.

The United States assured Israel that it would support Israel's rejection of the resolution—reaffirming a secret understanding that has allowed Israel to maintain a nuclear arsenal closed to international inspections, according to officials familiar with the arrangements. Again, the media were silent.

Indian officials greeted U.N. Resolution 1887 by announcing that India "can now build nuclear weapons with the same destructive power as those in the arsenals of the world's major nuclear powers," the *Financial Times* reported.

Both India and Pakistan are expanding their nuclear weapons programs. They have twice come dangerously close to nuclear war, and the problems that almost ignited this catastrophe are very much alive.

Obama greeted Resolution 1887 differently. The day before he was awarded the Nobel Prize for his inspiring commitment to peace, the Pentagon announced it was accelerating delivery of the most lethal non-nuclear weapons in the arsenal: thirteen-ton bombs for B-2 and B-52 stealth bombers, designed to destroy deeply hidden bunkers shielded by ten thousand pounds of reinforced concrete.

It's no secret the bunker busters are intended as a threat against Iran.

Planning for these "massive ordnance penetrators" began in the Bush years but languished until Obama called for developing them rapidly when he came into office.

Passed unanimously, Resolution 1887 calls for the end of threats of force and for all countries to join the NPT, as Iran did long ago. NPT non-signers are India, Israel and Pakistan, all of which developed nuclear weapons with U.S. help, in violation of the NPT. There is no thought of any of them joining the NPT. The only relevant threats of force are by the United States and Israel, against Iran.

Iran hasn't invaded another country for hundreds of years—unlike the United States, Israel and India (which occupies Kashmir, brutally).

The threat from Iran is minuscule. To believe Iran would use nuclear weapons to attack Israel, or anyone, "amounts to assuming that Iran's leaders are insane" and that they look forward to being reduced to "radioactive dust," strategic analyst Leonard Weiss observes, adding that Israel's missile-carrying submarines are "virtually impervious to preemptive military attack," not to speak of the immense U.S. arsenal.

In naval maneuvers in July [2009], Israel sent its Dolphin-class subs, capable of carrying nuclear missiles, through the Suez Canal and into the Red Sea, sometimes accompanied by warships, to a position from which they could attack Iran—as they have a "sovereign right" to do, according to U.S. Vice President Joe Biden.

Not for the first time, what is veiled in silence would receive front-page headlines in societies that valued their freedom and were concerned with the fate of the world.

The Iranian regime is harsh and repressive, and no

humane person wants Iran—or anyone else—to have nuclear weapons. But a little honesty would not hurt in addressing these problems.

The Nobel Peace Prize, of course, is not concerned
solely with reducing the threat of terminal nuclear war, but
rather with war generally, and the preparation for war. In
this regard, the selection of Obama raised eyebrows, not
least in Iran, surrounded by U.S. occupying armies.

On Iran's borders in Afghanistan and in Pakistan,
Obama has escalated Bush's war and is likely to proceed on
that course, perhaps sharply.

Obama has made clear that the United States intends
to retain a long-term major presence in the region. That
much is signaled by the huge city-within-a city called "the
Baghdad Embassy," unlike any embassy in the world.

Obama has announced the construction of mega-
embassies in Islamabad and Kabul, and huge consulates in
Peshawar and elsewhere.

Nonpartisan budget and security monitors report in
Government Executive that the "administration's request for
$538 billion for the Defense Department in fiscal 2010
and its stated intention to maintain a high level of funding
in the coming years put the president on track to spend
more on defense, in real dollars, than any other president
has in one term of office since World War II. And that's
not counting the additional $130 billion the administra-
tion is requesting to fund the wars in Iraq and Afghanistan
next year [2010], with even more war spending slated for
future years."

The Nobel Peace Prize committee might well have
made truly worthy choices, prominent among them the
remarkable Afghan activist Malalai Joya.

This brave woman survived the Russians, and then the radical Islamists whose brutality was so extreme that the population welcomed the Taliban. Joya has withstood the Taliban and now the return of the warlords under the Karzai government.

Throughout, Joya worked effectively for human rights, particularly for women; she was elected to parliament and then expelled when she continued to denounce warlord atrocities. She now lives underground under heavy protection, but she continues the struggle, in word and deed. By such actions, repeated by others everywhere as best they can, the prospects for peace edge closer to hopes.

The Legacy of 1989 in Two Hemispheres

DECEMBER 1, 2009

November [2009] marked the anniversary of major events in 1989: "the biggest year in world history since 1945," as British historian Timothy Garton Ash describes it.

That year "changed everything," Garton Ash writes. Mikhail Gorbachev's reforms within Russia and his "breathtaking renunciation of the use of force" led to the fall of the Berlin Wall on November 9—and to the liberation of Eastern Europe from Russian tyranny.

The accolades are deserved; the events, memorable. But alternative perspectives may be revealing.

German chancellor Angela Merkel provided such a perspective—unintentionally—when she called on all of us to "use this invaluable gift of freedom . . . to overcome the walls of our time."

One way to follow her good advice would be to dismantle the massive wall, dwarfing the Berlin wall in scale and length, now snaking through Palestinian territory in violation of international law.

The "annexation wall," as it should be called, is justified in terms of "security"—the default rationalization for state crimes. If security were the concern, the wall would be built along the border and made impregnable.

The purpose of this monstrosity, constructed with U.S. support and European complicity, is to allow Israel to take over valuable Palestinian land and significant water resources of the region, thus denying any viable national existence for the indigenous population of the former Palestine.

Another perspective on 1989 comes from Thomas Carothers, a scholar who served in "democracy enhancement" programs in the administration of former President Ronald Reagan.

After reviewing the record, Carothers concludes that all U.S. leaders have been "schizophrenic"—supporting democracy if it conforms to U.S. strategic and economic objectives, thus in Soviet satellites but not in U.S. client states.

This perspective is dramatically confirmed by the recent commemoration of the events of November 1989. The fall of the Berlin wall was rightly celebrated, but there was little notice of what happened one week later: on November 16, in El Salvador, the assassination of six leading Latin American intellectuals, Jesuit priests, along with their housekeeper and her daughter, by the elite, U.S.-armed Atlacatl battalion, fresh from renewed training at the JFK Special Warfare School in Fort Bragg, North Carolina.

The battalion and its cohorts had already compiled a bloody record through the grisly decade in El Salvador that began in 1980 with the assassination, by much the same hands, of Archbishop Oscar Romero, known as "the voice of the voiceless."

During the decade of the "war on terror" declared by the Reagan administration, the horror was similar throughout Central America. The reign of torture, murder and destruction in the region left hundreds of thousands dead.

The contrast between the liberation of Soviet satellites

and the crushing of hope in U.S. client states is striking and instructive—even more so when we broaden the perspective.

The assassination of the Jesuit intellectuals brought a virtual end to "liberation theology," the revival of Christianity that had its modern roots in the initiatives of Pope John XXIII and Vatican II, which he opened in 1962.

Vatican II "ushered in a new era in the history of the Catholic Church," theologian Hans Kung wrote. Latin American bishops adopted "the preferential option for the poor."

The bishops renewed the radical pacifism of the Gospels that had been put to rest when the Emperor Constantine established Christianity as the religion of the Roman Empire—"a revolution" that in less than a century converted "the persecuted church" to a "persecuting church," according to Kung.

In the post–Vatican II revival, Latin American priests, nuns and laypersons took the message of the Gospels to the poor and the persecuted, brought them together in base communities, and encouraged them to take their fate into their own hands.

Reaction to this heresy was violent repression. In the course of the terror and slaughter, the practitioners of liberation theology were a prime target.

Among them are the six martyrs of the church whose execution twenty years ago is now commemorated with a resounding silence, barely broken.

Last month [November 2009] in Berlin, the three presidents most involved in the fall of the Wall—George H.W. Bush, Mikhail Gorbachev and Helmut Kohl—discussed who deserves credit.

"I know now how heaven helped us," Kohl said. George H.W. Bush praised the East German people, who "for too long had been deprived of their God-given rights." Gorbachev suggested that the United States needs its own *perestroika*.

No doubts exist about responsibility for demolishing the attempt to revive the church of the Gospels in Latin America during the 1980s.

The School of the Americas (since renamed) in Fort Benning, Georgia, which trains Latin American officers, many with gruesome records, proudly announces that the U.S. Army helped to "defeat liberation theology"—assisted, to be sure, by the Vatican, using the gentler hand of expulsion and suppression.

The grim campaign to reverse the heresy set in motion by Vatican II received an incomparable literary expression in Dostoyevsky's parable of the Grand Inquisitor in *The Brothers Karamazov*.

In this tale, set in Seville at "the most terrible time of the Inquisition," Jesus Christ suddenly appears on the streets, "softly, unobserved, and yet, strange to say, everyone recognized him" and was "irresistibly drawn to him."

The Grand Inquisitor "bids the guards take Him and lead Him away" to prison. There he accuses Christ of coming to "hinder us" in the great work of destroying the subversive ideas of freedom and community. We follow not Thee, the Inquisitor admonishes Jesus, but rather Rome and "the sword of Caesar." We seek to be sole rulers of the earth so that we can teach the "weak and vile" multitude that "they will only become free when they renounce their freedom to us and submit to us." Then they will be timid and frightened and happy. So tomorrow, the Inquisitor says, "I must burn Thee."

Finally, however, the Inquisitor relents and releases "Him into the dark alleys of the town. The Prisoner went away."

The pupils of the U.S.-run School of the Americas practiced no such mercy.

Presidential "Peacekeeping"
in Latin America

DECEMBER 30, 2009

Barack Obama, the fourth U.S. president to win the Nobel Peace Prize, joins the others in the long tradition of peace-making so long as it serves U.S. interests.

All four presidents left their imprint on "our little region over here that has never bothered anybody," as U.S. Secretary of War Henry L. Stimson characterized the hemisphere in 1945.

Given the Obama administration's supportive stance toward the elections in Honduras in November (2009), it may be worthwhile to examine the record.

THEODORE ROOSEVELT

In his second term as president, Theodore Roosevelt said, "The expansion of the peoples of white, or European, blood during the past four centuries . . . has been fraught with lasting benefit to most of the peoples already dwelling in the lands over which the expansion took place," despite what Africans, Native Americans, Filipinos and other ben-eficiaries might mistakenly believe.

It was therefore "inevitable and in the highest de-gree desirable for the good of humanity at large, that the American people should ultimately crowd out the Mexi-cans" by conquering half of Mexico, and "It was out of the question to expect [Texans] to submit to the mastery of the weaker race."

Using gunboat diplomacy to steal Panama from Colombia to build the canal was also a gift to humanity.

WOODROW WILSON

Woodrow Wilson is the most honored of the presidential laureates and arguably the worst for Latin America.

Wilson's invasion of Haiti in 1915 killed thousands, restored virtual slavery and left much of the country in ruins.

Demonstrating his love of democracy, Wilson ordered his Marines to disband the Haitian parliament at gunpoint for failing to pass "progressive" legislation that allowed U.S. corporations to buy up the country. The problem was remedied when Haitians adopted a U.S.-written constitution, under Marine guns (with 99.9 percent approval, under 5 percent of the population participating). The achievement would be "beneficial to Haiti," Wilson's State Department assured its wards.

Wilson also invaded the Dominican Republic to ensure its welfare. Both countries were left under the rule of vicious national guards. Decades of torture, violence and misery there come down to us as a legacy of "Wilsonian idealism," a leading principle of U.S. foreign policy.

JIMMY CARTER

For President Jimmy Carter, human rights were "the soul of our foreign policy."

Robert Pastor, Carter's national security adviser for Latin America, explained some important distinctions between rights and policy: Regretfully, the administration had to support Nicaraguan dictator Anastasio Somoza's regime, and when that proved impossible, to maintain the U.S.-trained National Guard even after it had been massacring the population "with a brutality a nation usually reserves for its enemy," killing some forty thousand people.

As Pastor explained, the reason is elementary: "The United States did not want to control Nicaragua or the other nations of the region, but it also did not want developments to get out of control. It wanted Nicaraguans to act independently, *except* when doing so would affect U.S. interests adversely" (his emphasis).

BARACK OBAMA

President Barack Obama separated the United States from almost all of Latin America and Europe by accepting the military coup that overthrew Honduran democracy last June.

The coup reflected a "yawning political and socioeconomic divide," the *New York Times* reported. For the "small upper class," Honduran President Manuel Zelaya was becoming a threat to what they call "democracy," namely, the rule of "the most powerful business and political forces in the country."

Zelaya was initiating such dangerous measures as a rise in the minimum wage in a country where 60 percent live in poverty. He had to go.

Virtually alone, the United States recognized the November [2009] elections (with Pepe Lobo the victor) held under military rule—"a great celebration of democracy," according to Hugo Llorens, Obama's ambassador.

The endorsement also preserved the use of Honduras' Palmerola air base, increasingly valuable as the U. S. military is being driven out of most of Latin America.

After the elections, Lewis Anselem, Obama's representative to the Organization of American States, instructed the backward Latin Americans that they should recognize the military coup and join the United States "in the real world, not in the world of magical realism."

Obama broke new ground in supporting the military coup. The U.S. government funds the International Republican Institute (IRI) and the National Democratic Institute (NDI), which are supposed to promote democracy.

The IRI regularly supports military coups to overthrow elected governments, most recently in Venezuela in 2002 and Haiti in 2004.

But the NDI had held back. In Honduras, for the first time, Obama's NDI agreed to observe the elections under military rule, unlike the OAS and the United Nations, still wandering in the world of magical realism.

Given the close connections between the Pentagon and the Honduran military, and the enormous U.S. economic leverage in the country, it would have been a simple matter for Obama to join the Latin American–European effort to protect Honduran democracy.

But Obama preferred the traditional policy.

In his history of hemispheric relations, British scholar Gordon Connell-Smith writes, "While paying lip-service to the encouragement of representative democracy in Latin America, the United States has a strong interest in just the reverse," apart from "procedural democracy, especially the holding of elections, which only too often have proved farcical."

Functioning democracy may respond to popular concerns, while "the United States has been concerned with fostering the most favorable conditions for her private overseas investment."

It takes a large dose of what has sometimes been called "intentional ignorance" not to see the facts.

Such blindness must be guarded zealously if state violence is to proceed on course—always for the good of humanity, as Obama reminded us again in his Nobel Prize address.

The Corporate Takeover
of U.S. Democracy

FEBRUARY 1, 2010

January 21, 2010, will go down as a dark day in the history of U.S. democracy, and its decline.

On that day the U.S. Supreme Court ruled that the government may not ban corporations from political spending on elections—a decision that profoundly affects government policy, both domestic and international.

The decision heralds even further corporate takeover of the U.S. political system.

To the editors of the *New York Times*, the ruling "strikes at the heart of democracy" by having "paved the way for corporations to use their vast treasuries to overwhelm elections and intimidate elected officials into doing their bidding."

The court was split, 5-4, with the four reactionary judges (misleadingly called "conservative") joined by Justice Anthony M. Kennedy. Chief Justice John G. Roberts Jr. selected a case that could easily have been settled on narrow grounds and maneuvered the court into using it to push through a far-reaching decision that overturns a century of precedents restricting corporate contributions to federal campaigns.

Now corporate managers can in effect buy elections directly, bypassing more complex indirect means. It is well known that corporate contributions, sometimes packaged in complex ways, can tip the balance in elections, hence driving policy. The court has just handed much more power to the small sector of the population that dominates the economy.

Political economist Thomas Ferguson's "investment theory of politics" is a very successful predictor of government policy over a long period. The theory interprets elections as occasions on which segments of private sector power coalesce to invest to control the state.

The January 21 decision only reinforces the means to undermine functioning democracy.

The background is enlightening. In his dissent, Justice John Paul Stevens acknowledged that "we have long since held that corporations are covered by the First Amendment"—the constitutional guarantee of free speech, which would include support for political candidates.

In the early twentieth century, legal theorists and courts implemented the court's 1886 decision that corporations—these "collectivist legal entities"—have the same rights as persons of flesh and blood.

This attack on classical liberalism was sharply condemned by the vanishing breed of conservatives. Christopher G. Tiedeman described the principle as "a menace to the liberty of the individual, and to the stability of the American states as popular governments."

In his standard legal history, Morton Horwitz writes that the concept of corporate personhood evolved alongside the shift of power from shareholders to managers, and finally to the doctrine that "the powers of the board of directors . . . are identical with the powers of the corporation." In later years, corporate rights were expanded far beyond those of persons, notably by the mislabeled "free trade agreements." Under these agreements, for example, if General Motors establishes a plant in Mexico, it can demand to be treated just like a Mexican business ("national treatment")—quite unlike a Mexican of flesh and blood who might seek "national treatment" in New York, or even minimal human rights.

A century ago, Woodrow Wilson, then an academic, described an America in which "comparatively small groups of men," corporate managers, "wield a power and control over the wealth and the business operations of the country," becoming "rivals of the government itself."

In reality, these "small groups" increasingly have become government's masters. The Roberts court gives them even greater scope.

The January 21 decision came three days after another victory for wealth and power: the election of Republican candidate Scott Brown to replace the late Senator Edward M. Kennedy, the "liberal lion" of Massachusetts. Brown's election was depicted as a "populist upsurge" against the liberal elitists who run the government.

The voting data reveal a rather different story.

High turnouts in the wealthy suburbs, and low ones in largely Democratic urban areas, helped elect Brown. "Fifty-five percent of Republican voters said they were 'very interested' in the election," the *Wall Street Journal*–NBC poll reported, "compared with 38 percent of Democrats."

So the results were indeed an uprising against President Obama's policies: For the wealthy, he was not doing enough to enrich them further, while for the poorer sectors, he was doing too much to achieve that end.

The popular anger is quite understandable, given that

the banks are thriving, thanks to bailouts, while unemployment has risen to 10 percent.

In manufacturing, one in six is out of work—unemployment at the level of the Great Depression. With the increasing financialization of the economy and the hollowing out of productive industry, prospects are bleak for recovering the kinds of jobs that were lost.

Brown presented himself as the forty-first vote against health care—that is, the vote that could undermine majority rule in the U.S. Senate.

It is true that Obama's health care program was a factor in the Massachusetts election. The headlines are correct when they report that the public is turning against the program.

The poll figures help explain why: The bill does not go far enough. The *Wall Street Journal*–NBC poll found that a majority of voters disapprove of the handling of health care both by the Republicans and by Obama.

These figures align with recent nationwide polls. The public option was favored by 56 percent of those polled, and the Medicare buy-in at age fifty-five by 64 percent; both programs were quickly abandoned by Obama.

Eighty-five percent believe that the government should have the right to negotiate drug prices, as in other countries; Obama guaranteed Big Pharma that he would not pursue that option.

Large majorities favor cost-cutting, which makes good sense: U.S. per capita costs for health care are about twice those of other industrial countries, and health outcomes are at the low end.

But cost-cutting cannot be seriously undertaken when largesse is showered on the drug companies, and health care is in the hands of virtually unregulated private insurers—a costly and inefficient system peculiar to the United States.

The January 21 decision raises significant new barriers to overcoming the serious crisis of health care, or to addressing such critical issues as the looming environmental and energy crises. The gap between public opinion and public policy looms larger. And the damage to American democracy can hardly be overestimated.

The Unelected "Architects of Policy"

MARCH 10, 2010

Shifts in global power, ongoing or potential, are a lively topic among policymakers and observers. One question is whether (or when) China might displace the United States as the dominant global player, perhaps along with India.

Such a shift would return the global system to something like it was before the European conquests. Economic growth in China and India has been rapid, and thanks to their having rejected the West's policies of financial deregulation, they survived the recession better than most. Nonetheless, questions arise.

One standard measure of social health is the U.N. Human Development Index, available most recently through 2008. India ranks 134th, slightly above Cambodia and below Laos and Tajikistan, about where it has been for many years. China ranks 92nd—tied with Belize, a bit above Jordan, below the Dominican Republic and Iran.

India and China also have very high inequality, so more than a billion of their inhabitants fall far lower in the scale.

Another concern is the U.S. debt, which, it is feared, places the U.S. in thrall to China. Until recently, Japan has long been the biggest international holder of U.S. government debt. Creditor leverage, furthermore, is overrated, as the case of Japan illustrates.

In one dimension, military power, the United States

stands entirely alone. And Obama is setting new records with his military budget. Almost half the U.S. deficit is due to military spending, untouchable in the political system. The dysfunctional health care system, an international scandal, is the other major long-term contributor to the deficit. Economist Dean Baker has shown that if the United States had a health care system similar to that of other industrial powers, hardly a utopian idea, there would be no deficit and might be a surplus. But the power of financial institutions renders that too untouchable in the political system.

Nobel laureate Joseph Stiglitz and other economists warn that we should beware of "deficit fetishism." A deficit is a stimulus to recovery, and it can be overcome with a growing economy, as after World War II, when the deficit was far worse.

However, the framework of these discussions is misleading. The global system is not just an interaction among states, each pursuing some "national interest" abstracted from distribution of domestic power. That has long been understood.

Adam Smith concluded that the "principal architects" of policy in England were "merchants and manufacturers," who ensured that their own interests are "most peculiarly attended to," however "grievous" the effects on others, including the people of England.

Smith's maxim still holds, though today the "principal architects" are multinational corporations and particularly the financial institutions whose share in the economy has exploded since the 1970s.

In the United States we have just seen a dramatic illustration of the power of the financial institutions. In the last presidential election they provided the core of President Obama's funding.

Naturally they expected to be rewarded, and were—with the TARP (Troubled Asset Relief Program) bailouts, and a great deal more. Take Goldman Sachs, the top dog in the economy and political system. The firm made a mint by selling mortgage-backed securities and more complex financial instruments.

Aware of the flimsiness of the packages they were peddling, the firm took out bets with the insurance giant American International Group that the offerings would fail. When the financial system collapsed, AIG went down with it.

Goldman's architects of policy not only parlayed a bailout for Goldman itself but also arranged for taxpayers to save AIG from bankruptcy, thus rescuing Goldman.

Now Goldman is making record profits and paying out fat bonuses, and along with a few major banks, is bigger and more powerful than ever. The public is furious. People can see that the banks that were primary agents of the crisis are making out like bandits, while the population that rescued them is facing unemployment of nearly 10 percent.

Popular anger finally evoked a rhetorical shift from the administration, which responded with charges about greedy bankers, along with some policy suggestions that the financial industry doesn't like (the Volcker Rule and other proposals).

Since Obama was supposed to be their man in Washington, the principal architects wasted little time delivering their instructions: Unless Obama fell back into line, they would shift funds to the political opposition.

Within days, Obama informed the press that bankers

are fine "guys," singling out the chairs of the two biggest players, JP Morgan Chase and Goldman Sachs: "I, like most of the American people, don't begrudge people success or wealth. That's part of the free market system"—as "free markets" are interpreted in state capitalist doctrine.

That turnabout is a revealing snapshot of Smith's maxim in action.

The architects of policy are also at work on a real shift of power: from the global workforce to transnational capital.

Economist and China specialist Martin Hart-Landsberg explores the dynamic. China has become an assembly plant for a regional production system. Japan, Taiwan and other advanced Asian economies export high-tech parts and components to China, which assembles and exports the finished products.

The growing U.S. trade deficit with China has aroused concern. Less noticed is that the U.S. trade deficit with Japan and the rest of Asia has sharply declined as the new regional production system takes shape. U.S. manufacturers are following the same course, providing parts and components for China to assemble and export, mostly back to the United States. For the financial institutions, retail giants, and the owners and managers of manufacturing industries and sectors closely related to this nexus of power, these developments are heavenly.

And well understood. In 2007, Ralph Gomory, head of the Alfred P. Sloan Foundation, testified before Congress, "In this new era of globalization, the interests of companies and countries have diverged. In contrast with the past, what is good for America's global corporations is no longer necessarily good for the American people."

Consider IBM. By the end of 2008, more than 70 percent of IBM's workforce of four hundred thousand was abroad, *Business Week* reports. In 2009 IBM reduced its U.S. employment by another 8 percent. This case is particularly instructive in light of IBM's extensive reliance on innovation and development in the state sector, and direct subsidy by procurement and other means.

For the workforce, the outcome may be "grievous," in accordance with Smith's maxim, but it is fine for the principal architects of policy. Current research indicates that about one-fourth of U.S. jobs will be "offshorable" within two decades, and for those jobs that remain, security and decent pay will decline because of the increased competition from replaced workers.

This pattern follows thirty years of stagnation or decline for the majority as wealth poured into few pockets, leading to probably the highest inequality in U.S. history, much of it caused by the stratospheric rise in wealth of a fraction of one percent of the population.

While China is becoming the world's assembly plant and export platform, Chinese workers are suffering along with rest of the global workforce, as we would anticipate in a system designed to concentrate wealth and power and to set working people in competition with one another worldwide.

Globally, workers' share in national income has declined in many countries—dramatically so in China, leading to growing unrest in this highly inegalitarian society.

So we have another significant shift in global power: from the general population to the principal architects of the global system, a process aided by the undermining of functioning democracy in the most powerful states.

The future depends on how much the great majority is willing to endure, and whether a constructive response can develop that will confront the problems at the core of the state capitalist system of domination and control. If not, the results might be grim, as history more than amply reveals.

A "Regrettable" Event
in East Jerusalem

MARCH 29, 2010

Yet again the flashpoint is East Jerusalem, seized by Israel in the 1967 war—this time, a proposed 1,600-apartment complex in the Ramat Shlomo neighborhood. And yet again the aftermath has led to the death of Palestinians by Israeli gunfire.

On March 9 [2010], the Israeli interior ministry announced the new project during U.S. Vice President Joseph R. Biden's visit to Israel. President Obama had called for curbing settlement expansion in occupied territory.

Israeli Prime Minister Benjamin Netanyahu publicly apologized for the announcement's "regrettable" timing but insisted that Israel could build freely in East Jerusalem and elsewhere in the territories it intends to annex as well.

Biden had a private, angry exchange with Netanyahu, invoking U.S. military concern about the failure to resolve the Israeli-Palestinian conflict, according to the Israeli press.

"What you're doing here undermines the security of our troops who are fighting in Iraq, Afghanistan and Pakistan," Biden reportedly told Netanyahu. "That endangers us and it endangers regional peace."

On March 16 [2010], General David H. Petraeus, chief of the U.S. Central Command, voiced those concerns to the Senate Armed Services Committee: "The conflict foments anti-American sentiment due to a perception of U.S. favoritism for Israel."

A week later, Netanyahu and Obama met at the White House for talks later characterized as "contentious."

Netanyahu maintains a hard line on the settlements. And he makes no show of recognizing the viability of a Palestinian state. This intransigence reflects badly on U.S. credibility.

A similar, settlement-related contretemps flared up twenty years ago, leading President George H.W. Bush to impose limited sanctions on Israel in reaction to the brazen, insulting behavior of Prime Minister Yitzhak Shamir, who was quickly replaced. The Obama administration has made it clear that it would not take even the mild measures invoked by Bush senior.

The situation is now more serious. Within Israel, ultranationalist and religious sectors have risen with a narrow, parochial perspective. And U.S. forces are engaged in unpopular wars in the region.

Last May [2009], in Washington, Obama met with Netanyahu and Mahmoud Abbas, president of the Palestinian Authority. The meetings, and Obama's speech in Cairo in June [2009], have been interpreted as a turning point in U.S. Middle East policy.

A closer look, however, suggests reservations.

The U.S.-Israel interactions—with Abbas on the sidelines—hinged on two phrases: "Palestinian state" and "natural growth of settlements." Let's consider each in turn.

Obama has indeed pronounced the words "Palestinian state," echoing President George W. Bush. By contrast, the (unrevised) 1999 platform of Israel's governing party,

Netanyahu's Likud, "flatly rejects the establishment of a Palestinian Arab state west of the Jordan River."

It is also useful to recall that Netanyahu's 1996 government was the first in Israel to accept the possibility of a "Palestinian state," in an interesting way. The government agreed that Palestinians can call whatever fragments of Palestine are left to them "a state" if they like—or they can call them "fried chicken."

Last May [2009], Washington's position was presented most forcefully in U.S. Secretary of State Hillary Clinton's much-quoted statement rejecting "natural growth exceptions" to the official U.S. policy opposing new settlements.

Netanyahu and virtually the whole Israeli political spectrum insist on permitting such "natural growth," complaining that the U.S. is backing down from Bush's authorization of such expansion within his "vision" of a Palestinian state.

The Obama-Clinton formulation is not new. It repeats the wording of Bush's Road Map to a Palestinian State, which stipulates that in Phase I, Israel "freezes all settlement activity consistent with the [former U.S. Senator George J.] Mitchell report, including natural growth of settlement."

In Cairo, Obama kept to his familiar "blank-slate" style—with little substance but presented in a personable manner that allows listeners to write on the slate what they want to hear.

Obama echoed Bush's "vision" of a Palestinian state, without spelling out what he meant.

Obama said, "The United States does not accept the legitimacy of continued Israeli settlements." The operative words are "legitimacy" and "continued."

By omission, Obama indicated that he accepts Bush's "vision": The vast existing Israeli settlement and infrastructure projects on the West Bank are implicitly "legitimate," thus ensuring that the phrase "Palestinian state," referring to the scattered remnants in between, means "fried chicken."

Last November [2009], Netanyahu declared a ten-month suspension of new construction, with many exemptions, and entirely excluding Greater Jerusalem, where expropriation in Arab areas and construction for Jewish settlers, as at the Ramat Shlomo project, continues at a rapid pace.

These projects are doubly illegal: Like all settlements, they violate international law—and in Jerusalem, specific Security Council resolutions dating back to 1968.

In Jerusalem at the time, Hillary Clinton praised Netanyahu's "unprecedented" concessions on (illegal) construction, eliciting anger and ridicule in much of the world.

The Obama administration advocates a "reconceptualization" of the Middle East conflict, articulated most clearly last March by Senate Foreign Relations Committee chair John Kerry.

Israel is to be integrated among the "moderate" Arab states that are U.S. allies, confronting Iran and providing for U.S. domination of the vital energy-producing regions. Within that framework some unspecified Israel-Palestine settlement will find its place.

Meanwhile the bonds deepen between the United States and Israel. Close intelligence cooperation goes back over half a century.

U.S.-Israeli high-tech partnerships are flourishing. Intel, for example, is adding a gigantic installation to its

Kiryat Gat facility to implement a revolutionary reduction in the size of chips.

Ties between U.S. and Israeli military industry remain particularly close, so much so that Israel has been shifting development and manufacturing facilities to the United States, where access to U.S. military aid and markets is easier. Israel is also considering transfer of production of armored vehicles to the United States, over the objections of thousands of Israeli workers who will lose their jobs.

The relations also benefit U.S. military producers—doubly so, in fact, because supplies of U.S. government–funded weapons to Israel, which are themselves very profitable, also function as "teasers" that induce the rich Arab dictatorships ("moderates") to purchase great amounts of less sophisticated military equipment.

Israel also continues to provide the United States with a strategically located military base for pre-positioning weapons and other functions—most recently in January [2010], when the U.S. army moved to "double the value of emergency military equipment it stockpiles on Israeli soil," raising the level to $800 million.

"Missiles, armored vehicles, aerial ammunition and artillery ordnance are already stockpiled in the country," *Defense News* reports.

These are among the unparalleled services that Israel has been providing for U.S. militarism and global dominance, as well as for the U.S. high-tech economy.

They afford Israel a certain leeway to defy Washington's orders—though Israel is taking a big risk if it tries to push its luck, as history has repeatedly shown. The Ramat Shlomo arrogance clearly hit a nerve.

Israel can go only as far as the United States permits.

The United States has long been a direct participant even in Israeli crimes it formally condemns—but with a wink. It is up to the American people to determine whether the charade will continue.

NOTE

1. Adapted from *Hopes and Prospects* by Noam Chomsky (Haymarket Books, March 2010). Reprinted with permission.

Rust Belt Rage

APRIL 30, 2010

On February 18 [2010], Joe Stack, a fifty-three-year-old computer engineer, crashed his small plane into a building in Austin, Texas, hitting an IRS office, committing suicide, killing one other person and injuring others.

Stack left an anti-government manifesto explaining his actions. The story begins when he was a teenager living on a pittance in Harrisburg, Pennsylvania, near the heart of what was once a great industrial center.

His neighbor, in her eighties and surviving on cat food, was the "widowed wife of a retired steel worker. Her husband had worked all his life in the steel mills of central Pennsylvania with promises from big business and the union that, for his thirty years of service, he would have a pension and medical care to look forward to in his retirement.

"Instead he was one of the thousands who got nothing because the incompetent mill management and corrupt union (not to mention the government) raided their pension funds and stole their retirement. All she had was Social Security to live on."

He could have added that the super-rich and their political allies continue to try to take away Social Security, too.

Stack decided that he couldn't trust big business and would strike out on his own, only to discover that he also couldn't trust a government that cared nothing about people like him but only about the rich and privileged; or a legal system in which "there are two 'interpretations' for every law, one for the very rich, and one for the rest of us."

The government leaves us with "the joke we call the American medical system, including the drug and insurance companies [that] are murdering tens of thousands of people a year," with care rationed largely by wealth, not need.

Stack traces these ills to a social order in which "a handful of thugs and plunderers can commit unthinkable atrocities . . . and when it's time for their gravy train to crash under the weight of their gluttony and overwhelming stupidity, the force of the full federal government has no difficulty coming to their aid within days if not hours."

Stack's manifesto ends with two evocative sentences: "The communist creed: from each according to his ability, to each according to his need. The capitalist creed: from each according to his gullibility, to each according to his greed."

Poignant studies of the U.S. rust belt reveal comparable outrage among individuals who have been cast aside as state-corporate programs close plants and destroy families and communities.

An acute sense of betrayal comes readily to people who believed they had fulfilled their duty to society in a moral compact with business and government, only to discover they had been only instruments of profit and power.

Striking similarities exist in China, the world's second-largest economy, investigated by UCLA scholar Ching Kwan Lee.

Lee compared working-class outrage and desperation in the discarded industrial sectors of the United States and in what she calls China's rust belt—the state socialist industrial center in the Northeast, now abandoned for state capitalist development of the southeast sunbelt.

In both regions Lee found massive labor protests, but different in character. In the rust belt, workers express the same sense of betrayal as their U.S. counterparts—in their case, the betrayal of the Maoist principles of solidarity and dedication to development of the society that they thought had been a moral compact, only to discover that whatever it was, it is now bitter fraud.

Around the country, scores of millions workers dropped from work units "are plagued by a profound sense of insecurity," arousing "rage and desperation," Lee writes.

Lee's work and studies of the U.S. rust belt make clear that we should not underestimate the depth of moral indignation that lies behind the furious, often self-destructive bitterness about government and business power.

In the United States, the Tea Party movement—and even more so the broader circles it reaches—in part reflect the spirit of disenchantment. The Tea Party's anti-tax extremism is not as immediately suicidal as Joe Stack's protest, but it is suicidal nonetheless.

California today is a dramatic illustration. The world's greatest public system of higher education is being dismantled.

Governor Arnold Schwarzenegger says he'll have to eliminate state health and welfare programs unless the federal government forks over some $7 billion. Other governors are joining in.

Meanwhile a newly powerful states' rights movement is demanding that the federal government not intrude into our affairs—a nice illustration of what Orwell called "doublethink": the ability to hold two contradictory ideas in mind while believing both of them, practically a motto for our times.

California's plight results in large part from anti-tax fanaticism. It's much the same elsewhere, even in affluent suburbs.

Encouraging anti-tax sentiment has long been a staple of business propaganda. People must be indoctrinated to hate and fear the government, for good reasons: Of the existing power systems, the government is the one that in principle, and sometimes in fact, answers to the public and can constrain the depredations of private power.

However, anti-government propaganda must be nuanced. Business of course favors a powerful state that works for multinationals and financial institutions—and even bails them out when they destroy the economy.

But in a brilliant exercise in doublethink, people are led to hate and fear the deficit. That way, business's cohorts in Washington may agree to cut benefits and entitlements like Social Security (but not bailouts).

At the same time, people should not oppose what is largely creating the deficit—the growing military budget and the hopelessly inefficient privatized health care system.

It is easy to ridicule how Joe Stack and others like him articulate their concerns, but it's far more appropriate to understand what lies behind their perceptions and actions at a time when people with real grievances are being mobilized in ways that pose no slight danger to themselves and to others.

The Real Threat Aboard
the Freedom Flotilla

JUNE 4, 2010

Israel's violent attack on the Freedom Flotilla carrying humanitarian aid to Gaza shocked the world.

Hijacking boats in international waters and killing passengers is, of course, a serious crime.

But the crime is nothing new. For decades, Israel has been hijacking boats between Cyprus and Lebanon and killing or kidnapping passengers, sometimes holding them hostage in Israeli prisons.

Israel assumes that it can commit such crimes with impunity because the United States tolerates them and Europe generally follows the United States' lead.

As the editors of the *Guardian* rightly observed on June 1 [2010], "If an armed group of Somali pirates had yesterday boarded six vessels on the high seas, killing at least 10 passengers and injuring many more, a NATO task force would today be heading for the Somali coast." In this case, the NATO treaty obligates its members to come to the aid of a fellow NATO country—Turkey—attacked on the high seas.

Israel's pretext for the attack was that the Freedom Flotilla was bringing materials that Hamas could use for bunkers to fire rockets into Israel.

For many reasons, the pretext isn't credible—and even

if it were, would not justify international crimes. Israel can easily end the threat of rockets by peaceful means.

The background, briefly discussed earlier, is important. Hamas was designated a major terrorist threat particularly after it won a free election in January 2006. The United States and Israel sharply escalated their punishment of Palestinians, now for the crime of voting the wrong way.

The siege of Gaza, including a naval blockade, was a result. The siege intensified sharply in June 2007 after a civil war left Hamas in control of the territory.

What is commonly described as a Hamas military coup was in fact incited by the United States and Israel, in a crude attempt to overturn the elections that had brought Hamas to power.

That has been public knowledge at least since April 2008, when David Rose reported in *Vanity Fair* that George W. Bush, National Security Adviser Condoleezza Rice and her deputy, Elliott Abrams, "backed an armed force under Fatah strongman Muhammad Dahlan, touching off a bloody civil war in Gaza and leaving Hamas stronger than ever"—an account later confirmed by U.S. diplomat Norman Olsen, with long experience in the region.

Hamas terror included launching rockets into nearby Israeli towns—criminal, without a doubt, though only a minute fraction of routine U.S.-Israeli crimes in Gaza.

In June 2008, Israel and Hamas reached a cease-fire agreement. The Israeli government formally acknowledges that until Israel broke the agreement on November 4 of that year, invading Gaza and killing half a dozen Hamas activists, Hamas did not fire a single rocket, even though Israel had never adhered to the agreement, maintaining its siege.

Hamas offered to renew the cease-fire. The Israeli cabinet considered the offer and rejected it, preferring to launch its murderous invasion of Gaza on December 27 [2009].

Like other states, Israel has the right of self-defense. But did Israel have the right to use force in Gaza in the name of self-defense? International law, including the U.N. Charter, is unambiguous: A nation has such a right only if it has exhausted peaceful means. In this case such means were not even tried, although—or perhaps because—there was every reason to suppose that they would succeed.

Thus the invasion was sheer criminal aggression, and the same is true of Israel's resorting to force against the flotilla.

The siege is savage, designed to keep the caged animals barely alive so as to fend off international protest, but hardly more than that. It is the latest stage of long-standing Israeli plans, backed by the United States, to separate Gaza from the West Bank.

The Israeli journalist Amira Hass, a leading specialist on Gaza, outlines the history of the process of separation: "The restrictions on Palestinian movement that Israel introduced in January 1991 reversed a process that had been initiated in June 1967.

"Back then, and for the first time since 1948, a large portion of the Palestinian people again lived in the open territory of a single country—to be sure, one that was occupied, but was nevertheless whole."

Hass concludes: "The total separation of the Gaza Strip from the West Bank is one of the greatest achievements of Israeli politics, whose overarching objective is to prevent a solution based on international decisions and understandings and instead dictate an arrangement based on Israel's military superiority."

The Freedom Flotilla defied that policy, and so it must be crushed.

A framework for settling the Arab-Israeli conflict has existed since 1976, when the regional Arab States introduced a Security Council resolution calling for a two-state settlement on the international border, including all the security guarantees of U.N. Resolution 242, adopted after the June War in 1967.

The essential principles are supported by virtually the entire world, including the Arab League, the Organization of Islamic States (including Iran) and relevant non-state actors, including Hamas.

But the United States and Israel have led the rejection of such a settlement for three decades, with one crucial and highly informative exception of President Bill Clinton's last month in office, discussed earlier.

Today, the cruel legacy of a rejected peace lives on.

International law cannot be enforced against powerful states, except by their own citizens. That is always a difficult task, particularly when articulate opinion declares crime to be legitimate, either explicitly or by tacit adoption of a criminal framework—which is more insidious, because it renders the crimes invisible.

Storm Clouds over Iran

JULY 1, 2010

The dire threat of Iran is the most serious foreign policy crisis facing the Obama administration. Congress has just strengthened the sanctions against Iran, with even more severe penalties against foreign companies doing business there.

The administration has rapidly expanded U.S. offensive capacity in the African island of Diego Garcia, claimed by Britain, which had expelled the population so that the U.S. could build a massive base for attacking the Middle East and Central Asia.

The U.S. Navy reports sending a submarine tender to the island to service nuclear-powered submarines with Tomahawk missiles, which can carry nuclear warheads.

According to a U.S. Navy cargo manifest obtained by the *Sunday Herald* (Glasgow), the military equipment delivery to Diego Garcia includes 387 "bunker busters" for blasting hardened underground structures.

"They are gearing up totally for the destruction of Iran," Dan Plesch, director of the Center for International Studies and Diplomacy at the University of London, told the *Sunday Herald*. "U.S. bombers and long-range missiles are ready today to destroy 10,000 targets in Iran in a few hours."

The Arab press reports that an American fleet (with an Israeli vessel) has just passed through the Suez Canal on the way to the Persian Gulf, where its task is "to implement the sanctions against Iran and supervise the ships going to and from Iran."

British and Israeli media report that Saudi Arabia is providing a corridor for Israeli bombing of Iran (denied by Saudi Arabia).

On his return from Afghanistan to reassure NATO allies after the shift of command following General Stanley A. McChrystal's resignation, Admiral Michael Mullen, U.S. Chairman of the Joint Chiefs of Staff, visited Israel to meet Israel Defense Forces Chief of Staff Gabi Ashkenazi, continuing an annual strategic dialogue.

The meeting focused "on the preparation by both Israel and the U.S. for the possibility of a nuclear-capable Iran," according to *Haaretz*, which reports further that Mullen emphasized, "I always try to see challenges from Israeli perspective."

Some respected analysts describe the Iranian threat in apocalyptic terms. Amitai Etzioni warns, "The U.S. will have to confront Iran or give up the Middle East." If Iran's nuclear program proceeds, he asserts, Turkey, Saudi Arabia and other states will "move toward" the new Iranian "superpower." In less fevered rhetoric, a regional alliance might take shape independent of the United States.

In the U.S. Army journal *Military Review*, Etzioni urges a U.S. attack that targets not only Iran's nuclear facilities but also its non-nuclear military assets, including infrastructure—meaning the civilian society. "This kind of military action is akin to sanctions—causing 'pain' in order to change behavior, albeit by much more powerful means," he writes.

An authoritative analysis of the Iranian threat is provided by a U.S. Department of Defense report presented to Congress (in classified and unclassified form) in April [2010].

Iran's military spending is "relatively low compared to the rest of the region," the report maintains. Iran's military doctrine is strictly "defensive . . . designed to slow an invasion and force a diplomatic solution to hostilities." With regard to the nuclear option, "Iran's nuclear program and its willingness to keep open the possibility of developing nuclear weapons [are] a central part of its deterrent strategy."

To Washington, Iranian deterrent capacity is an illegitimate exercise of sovereignty that interferes with U.S. global designs. Specifically, it threatens U.S. control of Middle East energy resources.

But Iran's threat goes beyond deterrence. Iran is also seeking to expand its influence in the region, which is seen as "destabilization," presumably by contrast with the "stabilizing" U.S. invasion and military occupation of Iran's neighbors.

Beyond these crimes, the study continues, Iran is also supporting terrorism by backing Hezbollah and Hamas, the major political forces in Lebanon and in Palestine (if elections count).

The model for democracy in the Muslim world, despite serious flaws, is Turkey, which has relatively free elections.

The Obama administration was incensed when Turkey joined with Brazil in arranging with Iran to restrict its enrichment of uranium. The United States quickly undermined the deal by ramming through a U.N. Security Council resolution with new sanctions against Iran that were so meaningless that China cheerfully joined at once—recognizing that at most the sanctions would impede Western interests in competing with China for Iran's resources.

Not surprisingly, Turkey (along with Brazil) voted against the U.S. sanctions motion in the Security Council. The other regional member, Lebanon, abstained.

These actions aroused further consternation in Washington. Philip Gordon, the Obama administration's top diplomat on European affairs, warned Turkey that its actions are not understood in the United States and that it must "demonstrate its commitment to partnership with the West," the Associated Press reported, calling it "a rare admonishment of a crucial NATO ally."

The political class understands as well. Steven A. Cook, a scholar with the Council on Foreign Relations, observed that the critical question now is "How do we keep the Turks in their lane?"—following orders like good democrats.

There is no indication that other countries in the region favor U.S. sanctions any more than Turkey does. Pakistan and Iran, meeting in Turkey, recently signed an agreement for a new pipeline. Even more worrisome for the United States is that the pipeline might extend to India.

The 2008 U.S. treaty with India supporting its nuclear programs was intended to stop India from joining the pipeline, according to Moeed Yusuf, a South Asia adviser to the United States Institute of Peace, expressing a common interpretation.

India and Pakistan are two of the three nuclear powers that have refused to sign the Non-Proliferation Treaty, the third being Israel. All have developed nuclear weapons with U.S. support, and still do.

No sane person wants Iran, or any nation, to develop nuclear weapons. One obvious way to mitigate or eliminate this threat is to establish a Middle East nuclear-weapons-free zone.

The issue arose (again) at the NPT conference at United Nations headquarters in early May [2010]. Egypt, as chair of the 118 nations of the Non-Aligned Movement, proposed that the conference back a plan calling for the start of negotiations in 2011 on a Middle East nuclear-weapons-free zone, as had been agreed by the West, including the United States, at the 1995 review conference on the NPT.

Washington still formally agrees, but insists that Israel be exempted—and has given no hint of allowing such provisions to apply to the United States.

Instead of taking practical steps toward reducing the nightmarish threat of nuclear weapons proliferation in Iran or elsewhere, the United States is moving to reinforce control of the vital Middle Eastern oil-producing regions, by violence if other means do not succeed.

The War in Afghanistan:
Echoes of Vietnam

AUGUST 1, 2010

The War Logs—a six-year archive of classified military documents about the war in Afghanistan, released on the Internet by the organization WikiLeaks—documents a grim struggle becoming grimmer, from the U.S. perspective. And for the Afghans, a mounting horror.

The War Logs, however valuable, may contribute to the unfortunate and prevailing doctrine that wars are wrong only if they aren't successful—rather like the Nazis felt after Stalingrad.

Last month [July 2010] came the fiasco of General Stanley A. McChrystal, forced to retire as commander of U.S. forces in Afghanistan and replaced by his superior, General David H. Petraeus.

A plausible consequence is a relaxation of the rules of engagement so that it becomes easier to kill civilians, and an extension of the war well into the future as Petraeus uses his clout in Congress to achieve this result.

Afghanistan is President Obama's principal current war. The official goal is to protect ourselves from al-Qaida, a virtual organization, with no specific base—a "network of networks" and "leaderless resistance," as it's been called in the professional literature. Now, even more so than before, al-Qaida consists of relatively independent factions, loosely associated throughout the world.

The CIA estimates that fifty to one hundred al-Qaida activists may now be in Afghanistan, and there is no indication that the Taliban want to repeat the mistake of offering sanctuary to al-Qaida.

By contrast, the Taliban appear to be well established in their vast forbidding landscape, including a large part of the Pashtun territories.

In February [2010], in the first exercise of Obama's new strategy, U.S. Marines conquered Marja, a minor district in Helmand province, the main center of the insurgency.

There, reported the *New York Times*' Richard A. Oppel Jr., "the Marines have collided with a Taliban identity so dominant that the movement appears more akin to the only political organization in a one-party town, with an influence that touches everyone. . . .

"'We've got to re-evaluate our definition of the word 'enemy,' said Brigadier General Larry Nicholson, commander of the Marine expeditionary brigade in Helmand Province. 'Most people here identify themselves as Taliban. . . . We have to readjust our thinking so we're not trying to chase the Taliban out of Marja, we're trying to chase the enemy out.'"

The Marines are facing a problem that has always bedeviled conquerors, one that is very familiar to the United States from Vietnam. In 1966, Douglas Pike, the leading U.S. government scholar on Vietnam, lamented that the enemy—the National Liberation Front [NLF]—was the only "truly mass-based political party in South Vietnam."

Any effort to compete with that enemy politically would be like a conflict between a minnow and a whale, Pike recognized. We therefore had to overcome the NLF's political force by using our comparative advantage, violence—with horrifying results.

Others have faced similar problems: for example, the Russians in Afghanistan during the 1980s, where they won every battle but lost the war.

Writing of another U.S. invasion—the Philippines in 1898—Bruce Cumings, an Asia historian at the University of Chicago, made an observation that applies all too aptly to Afghanistan today: "When a sailor sees that his heading is disastrous he changes course, but imperial armies sink their boots in quicksand and keep marching, if only in a circle, while the politicians plum the phrase book of American ideals."

After the Marja triumph, the U.S.-led forces were expected to assault the major city of Kandahar, where, according to a U.S. Army poll in April [2010], the military operation is opposed by 95 percent of the population, and five out of six regard the Taliban as "our Afghan brothers"—again, echoes of earlier conquests. The Kandahar plans were delayed, part of the background for McChrystal's leave-taking.

Under these circumstances, it is not surprising that U.S. authorities are concerned that domestic public support for the war in Afghanistan may erode even further.

In May [2010], WikiLeaks released a March [2010] CIA memorandum about how to sustain Western Europe's support for the war. The memorandum's subtitle: "Why Counting on Apathy Might Not Be Enough."

"The Afghanistan mission's low public salience has allowed French and German leaders to disregard popular opposition and steadily increase their troop contributions to the International Security Assistance Force [ISAF]," the memorandum states.

"Berlin and Paris currently maintain the third and fourth highest ISAF troop levels, despite the opposition of 80 percent of German and French respondents to increased ISAF deployments." It is therefore necessary to "tailor messaging" to "forestall or at least contain backlash."

For France, the CIA recommends propaganda designed to address "acute French concern for civilians and refugees" and to elicit French guilt for abandoning them, in particular stressing girls' education, which can become "a rallying point for France's largely secular public, and give voters a reason to support a good and necessary cause despite casualties." The facts are as usual irrelevant. For example, the advances in girls' education in Kabul under Russians, or the actual impact of the military operations.

The CIA memorandum should remind us that states have an internal enemy: their own population, which must be controlled when state policy is opposed by the public.

Democratic societies rely not on force but on propaganda, engineering consent by "necessary illusion" and "emotionally potent oversimplification," to quote the recommendations of Obama's favorite philosopher, Reinhold Niebuhr.

The battle to control the internal enemy, as always, remains highly pertinent—indeed, the future of the war in Afghanistan may hinge on it.

China and the New World Order, Part 1

SEPTEMBER 1, 2010

Amid all the alleged threats to the world's reigning super-power, one rival is quietly, forcefully emerging: China. And the U.S. is closely scrutinizing China's intentions.

On August 13 [2010], a Pentagon study expressed concern that China is expanding its military forces in ways that "could deny the ability of American warships to operate in international waters off the coast," Thom Shanker reports in the *New York Times*.

Washington is alarmed that "China's lack of openness about the growth, capabilities and intentions of its military injects instability to a vital region of the globe."

The United States, on the other hand, is quite open about its intention to operate freely throughout the "vital region of the globe" surrounding China (as elsewhere).

The United States advertises its vast capacity to do so with a growing military budget that roughly matches that of the rest of the world combined, hundreds of military bases across the globe, and a huge lead in the technology of destruction and domination.

China's lack of understanding of the rules of international civility was illustrated by its objections to the plan for the advanced nuclear-powered aircraft carrier *USS George Washington* to take part in the U.S.-South Korea military exercises near China's coast in July [2010], with the alleged capacity to strike Beijing.

By contrast, the West understands that such U.S. operations are all undertaken to defend stability and its own security.

The term "stability" has a technical meaning in discourse on international affairs: domination by the United States. Thus no eyebrows are raised when James Chace, former editor of *Foreign Affairs*, explains that in order to achieve "stability" in Chile in 1973, it was necessary to "destabilize" the country—by overthrowing the elected government of President Salvador Allende and installing the dictatorship of General Augusto Pinochet, which proceeded to slaughter and torture with abandon and to set up a terror network that helped install similar regimes elsewhere, with U.S. backing, in the interest of stability and security.

It is routine to recognize that U.S. security requires absolute control. The premise was given a scholarly imprimatur by historian John Lewis Gaddis of Yale University in "Surprise, Security, and the American Experience," in which he investigates the roots of President George W. Bush's preventive war doctrine.

The operative principle is that expansion is "the path to security," a doctrine that Gaddis admiringly traces back almost two centuries—to President John Quincy Adams, the intellectual author of Manifest Destiny.

When Bush warned "that Americans must 'be ready for preemptive action when necessary to defend our liberty and to defend our lives,'" Gaddis observes, "he was echoing an old tradition rather than establishing a new one," reiterating principles that presidents from Adams to Woodrow Wilson "would all have understood . . . very well."

Likewise Wilson's successors, to the present. President Bill Clinton's doctrine was that the United States is entitled to use military force to ensure "uninhibited access to key markets, energy supplies and strategic resources," with no need even to concoct pretexts of the Bush II variety.

According to Clinton's defense secretary, William Cohen, the United States therefore must keep huge military forces "forward deployed" in Europe and Asia "in order to shape people's opinions about us" and "to shape events that will affect our livelihood and our security." This prescription for permanent war is a new strategic doctrine, military historian Andrew Bacevich observes, later amplified by Bush II and President Barack Obama.

As every Mafia don knows, even the slightest loss of control might lead to an unraveling of the system of domination as others are encouraged to follow a similar path.

This central principle of power is formulated as the "domino theory," in the language of policymakers, which translates in practice to the recognition that the "virus" of successful independent development might "spread contagion" elsewhere, and therefore must be destroyed while potential plague victims are inoculated, usually by brutal dictatorships.

According to the Pentagon study, China's military budget expanded to an estimated $150 billion in 2009, approaching "one-fifth of what the Pentagon spent to operate and carry out the wars in Iraq and Afghanistan" in that year, only a fraction of the total U.S. military budget, of course.

The United States' concerns are understandable, if one takes into account the virtually unchallenged assumption that the United States must maintain "unquestioned power" over much of the world, with "military and economic supremacy," while ensuring the "limitation of any exercise of sovereignty" by states that might interfere with its global designs.

These were the principles formulated by high-level planners and foreign policy experts during World War II, as they developed the framework for the postwar world, which was largely implemented.

The United States was to maintain this dominance in a "Grand Area," which was to include at a minimum the Western Hemisphere, the Far East and the former British empire, including the crucial energy resources of the Middle East.

As Russia began to grind down Nazi armies after Stalingrad, Grand Area goals extended to as much of Eurasia as possible. It was always understood that Europe might choose to follow an independent course—perhaps the Gaullist vision of a Europe from the Atlantic to the Urals. The North Atlantic Treaty Organization was partially intended to counter this threat, and the issue remains very much alive today as NATO is expanded to a U.S.-run intervention force responsible for controlling the "crucial infrastructure" of the global energy system on which the West relies.

Since becoming the world-dominant power during World War II, the United States has sought to maintain a system of global control. But that project is not easy to sustain. The system is visibly eroding, with significant implications for the future. China is an increasingly influential player—and challenger.

That common theme about today's world might remind us about the early days of "American decline," another common theme of today. That decline began long ago. The United States was at the peak of its power in 1945. The first major blow to its plans for global dominance was in 1949, when "the People's Republic of China" declared independence, an event that is conventionally called "the loss of China"—an interesting phrase; one can only lose what one possesses. That was the first of many "losses." China's current challenge is of a different order.

China and the New World Order, Part 2

OCTOBER I, 2010

Of all the "threats" to world order, one of the most persistent is democracy, unless it is under imperial control, and more generally, the assertion of independence. Such fears have guided imperial power throughout history.

In South America, Washington's traditional backyard, the subjects are increasingly disobedient. Their steps toward independence advanced further in February [2010] with the formation of the Community of Latin American and Caribbean States, which includes all states in the hemisphere apart from the United States and Canada.

For the first time since the Spanish and Portuguese conquests five hundred years ago, South America is moving toward integration, a prerequisite to independence. It is also beginning to address the internal scandal of a continent that is endowed with rich resources but dominated by tiny islands of wealthy elites in a sea of misery.

Furthermore, South-South relations are developing, with China playing a leading role, both as a consumer of raw materials and as an investor. Its influence is growing rapidly and has surpassed the United States' in some resource-rich countries.

More significant still are changes in the Middle Eastern arena. Sixty years ago, the influential planner A.A. Berle advised that controlling the region's incomparable energy resources would yield "substantial control of the world."

Correspondingly, loss of control would threaten the project of global dominance. By the 1970s, the major producers nationalized their hydrocarbon reserves, but the West retained substantial influence. In 1979, Iran was "lost" with the overthrow of the shah's dictatorship, which had been imposed by a U.S.-U.K. military coup in 1953 to ensure that this prize would remain in the proper hands.

By now, however, control is slipping away even among the traditional U.S. clients.

The largest known hydrocarbon reserves are in Saudi Arabia, a U.S. dependency ever since the United States displaced Britain there in a mini-war conducted during World War II. The United States remains by far the largest investor in Saudi Arabia and its major trading partner, and Saudi Arabia helps support the U.S. economy via investments.

However, more than half of Saudi oil exports now go to Asia, and its plans for growth face east. The same might turn out to be true of Iraq, the country with the second-largest reserves, if it can rebuild from the massive destruction of the murderous U.S.-U.K. sanctions and the invasion. And U.S. policies are driving Iran, the third major producer, in the same direction.

China is now the largest importer of Middle Eastern oil and the largest exporter to the region, replacing the United States. Trade relations are growing fast, doubling in the past five years.

The implications for world order are significant, as is the quiet rise of the Shanghai Cooperation Organization, which includes much of Asia but has banned the United States—potentially "a new energy cartel involving both producers and consumers," observes economist Stephen King, author of *Losing Control: The Emerging Threats to Western Prosperity*.

In Western policy-making circles and among political commentators, 2010 is called "the year of Iran." The Iranian threat is considered to pose the greatest danger to world order and to be the primary focus of U.S. foreign policy, with Europe trailing along politely as usual. It is officially recognized that the threat is not military: Rather, it is the threat of independence.

To maintain "stability" the U.S. has imposed harsh sanctions on Iran, but outside of Europe, few are paying attention. The nonaligned countries—most of the world— have strongly opposed U.S. policy toward Iran for years.

Nearby Turkey and Pakistan are constructing new pipelines to Iran, and trade is increasing. Arab public opinion is so enraged by Western policies that a majority even favors Iran's development of nuclear weapons.

The conflict benefits China. "China's investors and traders are now filling a vacuum in Iran as businesses from many other nations, especially in Europe, pull out," Clayton Jones reports in the *Christian Science Monitor*. In particular, China is expanding its dominant role in Iran's energy industries.

Washington is reacting with a touch of desperation. In August [2010], the State Department warned that "If China wants to do business around the world it will also have to protect its own reputation, and if you acquire a reputation as a country that is willing to skirt and evade international responsibilities that will have a long-term impact . . . their international responsibilities are clear"— namely, to follow U.S. orders.

Chinese leaders are unlikely to be impressed by such talk, the language of an imperial power desperately trying to cling to authority it no longer has. A far greater threat to imperial dominance than Iran is China's refusing to obey orders—and indeed, as a major and growing power, dismissing them with contempt, a stand with deep historical resonance.

The U.S. Elections: Outrage, Misguided

NOVEMBER 4, 2010

The U.S. midterm elections register a level of anger, fear and disillusionment in the country like nothing I can recall in my lifetime. Since the Democrats are in power, they bear the brunt of the revulsion over our current socioeconomic and political situation.

More than half the "mainstream Americans" in a Rasmussen poll last month [October 2010] said they view the Tea Party movement favorably—a reflection of the spirit of disenchantment.

The grievances are legitimate. For more than thirty years, real incomes for the majority of the population have stagnated or declined while work hours and insecurity have increased, along with debt. Wealth has accumulated, but in very few pockets, leading to unprecedented inequality.

These consequences mainly spring from the financialization of the economy since the 1970s and the corresponding hollowing-out of domestic production, both developments related to the decline in the rate of profit in manufacturing, as historian and political economist Robert Brenner has analyzed in detail. Spurring the process is the deregulation mania favored by Wall Street and supported by economists mesmerized by efficient-market myths, and other decisions flowing from the concentration of economic and political power.

People see that the bankers who were largely responsible for the financial crisis and who were saved from bankruptcy by the public are now reveling in record profits and huge bonuses. Meanwhile official unemployment stays at about 10 percent. Manufacturing is at Depression levels: one in six out of work, with good jobs unlikely to return.

People rightly want answers, and they are not getting them except from voices that tell tales that have some internal coherence—if you suspend disbelief and enter into their world of irrationality and deceit.

Ridiculing Tea Party shenanigans is a serious error, however. It is far more appropriate to understand what lies behind the movement's popular appeal, and to ask ourselves why justly angry people are being mobilized by the extreme right and not by the kind of constructive activism that rose during the Depression, like organizing the CIO (Congress of Industrial Organizations).

Now Tea Party sympathizers are hearing that every institution—government, corporations and the professions—is rotten, and that nothing works.

Amid the joblessness and foreclosures, the Democrats can't complain about the policies that led to the disaster. President Ronald Reagan and his Republican successors may have been the worst culprits, but the policies began with President Jimmy Carter and accelerated under President Bill Clinton. During the presidential election [2008], Barack Obama's primary constituency was financial institutions, which have gained remarkable dominance over the economy in the past generation.

That incorrigible eighteenth-century radical Adam Smith, speaking of England, observed that the principal architects of power were the owners of the society—in his day the merchants and manufacturers—and they made sure that government policy would attend scrupulously to their interests, however "grievous" the impact on the people of England; and worse, on the victims of "the savage injustice of the Europeans" abroad.

A modern and more sophisticated version of Smith's maxim is political economist Thomas Ferguson's "investment theory of politics," which sees elections as occasions when groups of investors coalesce in order to control the state by selecting the architects of policies who will serve their interests.

Ferguson's theory turns out to be a very good predictor of policy over long periods. That should hardly be surprising. Concentrations of economic power will naturally seek to extend their sway over any political process. The dynamic happens to be extreme in the United States.

The corporate high rollers have a defense against charges of "greed" and disregard for the health of the society. Their task is to maximize profit and market share; in fact, that's their legal obligation. If they don't fulfill that mandate, they'll be replaced by someone who will. They also ignore systemic risk: the likelihood that their transactions will harm the economy generally. Such "externalities" are not their concern—not because they are bad people, but for institutional reasons.

When the bubble bursts, the risk takers can flee to the shelter of the nanny state that they nourish. Bailouts—a kind of government insurance policy—are among many perverse incentives that magnify market inefficiencies.

"There is growing recognition that our financial

system is running a doomsday cycle," economists Peter Boone and Simon Johnson wrote in the *Financial Times* in January [2010]. "Whenever it fails, we rely on lax money and fiscal policies to bail it out. This response teaches the financial sector: Take large gambles to get paid handsomely, and don't worry about the costs—they will be paid by taxpayers" through bailouts and other devices, and the financial system "is thus resurrected to gamble again—and to fail again."

The doomsday metaphor applies more ominously outside the financial world. The American Petroleum Institute, backed by the Chamber of Commerce and other business lobbies, has intensified its efforts to persuade the public to dismiss concerns about anthropogenic global warming—with some success, polls indicate. Among Republican congressional candidates in the 2010 election, virtually all reject global warming.

The executives behind the propaganda no doubt know that global warming is real, and prospects grim. But the fate of the species is an externality that the executives must ignore, to the extent that market systems prevail. And there will be no one to ride to the rescue if the worst-case scenario unfolds.

I am just old enough to remember those chilling and terrifying days of Germany's descent from decency to Nazi barbarism, to borrow the words of Fritz Stern, the distinguished scholar of German history. In a 2005 article, Stern indicates that he has the future of the United States in mind when he reviews "a historic process in which resentment against a disenchanted secular world found deliverance in the ecstatic escape of unreason."

The world is too complex for history to repeat, but there are nevertheless lessons to keep in mind as we register the consequences of another election cycle. No shortage of tasks waits for those who seek to present an alternative to misguided rage and indignation, helping to organize the countless disaffected and to lead the way to a better future.

The Charade of Israeli-Palestinian Talks

DECEMBER 1, 2010

Washington's pathetic capitulation to Israel while pleading for a meaningless three-month freeze on settlement expansion—excluding Arab East Jerusalem—should go down as one of the most humiliating moments in U.S. diplomatic history.

In September [2010] the last (limited) settlement freeze ended, leading the Palestinians to cease direct talks with Israel. Now the Obama administration, desperate to lure Israel into a new freeze and thus revive the talks, is grasping at invisible straws—and lavishing gifts on a far-right Israeli government.

The gifts include $3 billion for fighter jets. The largesse also happens to be another taxpayer grant to the U.S. arms industry, which gains doubly from programs to expand the militarization of the Middle East.

U.S. arms manufacturers are subsidized not only to develop and produce advanced equipment for a state that is virtually part of the U.S. military-intelligence establishment but also to provide second-rate military equipment to the Gulf states—currently a precedent-breaking $60 billion arms sale to Saudi Arabia, which is a transaction that also recycles petrodollars to an ailing U.S. economy.

Israeli and U.S. high-tech civilian industries are closely integrated. It is small wonder that the most fervent support for Israeli actions comes from the business press and the Republican Party, the more extreme of the two business-oriented political parties. The pretext for the huge arms sales to Saudi Arabia is defense against the "Iranian threat."

However, the Iranian threat is not military, as the Pentagon and U.S. intelligence have emphasized. Were Iran to develop a nuclear weapons capacity, the purpose would be deterrent—presumably to ward off a U.S.-Israeli attack.

The real threat, in Washington's view, is that Iran is seeking to expand its influence in neighboring countries "stabilized" by U.S. invasion and occupation.

The official line is that the Arab states are pleading for U.S. military aid to defend themselves against Iran. True or false, the claim provides interesting insight into the reigning concept of democracy. Whatever the ruling dictatorships may prefer, Arabs in a recent poll released by the Brookings Institute rank the major threats to the region as Israel (88 percent), the United States (77 percent) and Iran (10 percent).

It is interesting that virtually all discussion by U.S. officials and commentators on the revelations of the just-released WikiLeaks cables totally ignored Arab public opinion, keeping to the views of the reigning dictators, often expressing considerable euphoria that "the Arabs support us" on Iran.

The U.S. gifts to Israel also include diplomatic support. According to current reports, Washington pledges to veto any U.N. Security Council actions that might annoy Israel's leaders and to drop any call for further extension of a settlement freeze.

Hence, by agreeing to the three-month pause, Israel

will no longer be disturbed by the paymaster as it expands its criminal actions in the occupied territories.

That these actions are criminal has not been in doubt in Israel since late 1967, when Israel's leading legal authority, international jurist Theodor Meron, advised the government that its plans to initiate settlements in the occupied territories violated the Fourth Geneva Convention, a core principle of international humanitarian law, established in 1949 to criminalize the horrors of the Nazi regime.

Meron's conclusion was endorsed by Justice Minister Ya'akov Shimson Shapira, and shortly after by Defense Minister Moshe Dayan, historian Gershom Gorenberg relates in his book *The Accidental Empire*.

Dayan informed his fellow ministers, "We must consolidate our hold so that over time we will succeed in 'digesting' Judea and Samaria [the West Bank] and merging them with 'little' Israel," meanwhile "dismember[ing] the territorial contiguity" of the West Bank, all under the usual pretense "that the step is necessary for military purposes."

Dayan had no doubts, or qualms, about what he was recommending: "Settling Israelis in occupied territory contravenes, as is known, international conventions," he observed. "But there is nothing essentially new in that."

Dayan's correct assumption was that the boss in Washington might object formally, but with a wink, and would continue to provide the decisive military, economic and diplomatic support for the criminal endeavors.

The criminality has been underscored by repeated Security Council resolutions, more recently by the International Court of Justice, with the basic agreement of U.S. Justice Thomas Buergenthal in a separate declaration. Israel's actions also violate U.N. Security Council resolutions concerning Jerusalem. But everything is fine as long as Washington winks.

Back in Washington, the Republican super-hawks are even more fervent in their support for Israeli crimes. Eric Cantor, the new majority leader in the House of Representatives, "has floated a novel solution to protect aid for Israel from the current foreign aid backlash," Glenn Kessler reports in the *Washington Post*: "giving the Jewish state its own funding account, thus removing it from funds for the rest of the world."

The issue of settlement expansion is simply a diversion. The real issue is the existence of the settlements and related infrastructure developments. These have been carefully designed so that Israel has already taken over more than 40 percent of the occupied West Bank, including suburbs of Jerusalem and Tel Aviv; the arable land; and primary water sources of the region on the Israeli side of the Separation Wall—in reality an annexation wall.

Since 1967, Israel has vastly expanded the borders of Jerusalem in violation of Security Council orders and despite universal international objection (including from the United States, at that time).

The focus on settlement expansion, and Washington's groveling, are not the only farcical elements of the current negotiations. The very structure is a charade. The United States is portrayed as an "honest broker" seeking to mediate between two recalcitrant adversaries. But serious negotiations would be conducted by some neutral party, with the United States and Israel on one side, and the world on the other.

It is hardly a secret that for thirty-five years the United States and Israel have stood virtually alone in opposition to a consensus on a political settlement that is close to universal. With brief and rare departures, the two rejectionist states have preferred illegal expansion to security. Unless Washington's stand changes, political settlement is effectively barred. And expansion, with its reverberations throughout the region and the world, continues.

Breaking the Israel-Palestine Deadlock

JANUARY 1, 2011

While intensively engaged in illegal settlement expansion, the government of Israel is also seeking to deal with two problems: a global campaign of what it perceives as "delegitimation"—that is, objections to its crimes and withdrawal of participation in them—and a parallel campaign of legitimation of Palestine.

The "delegitimation," which is progressing rapidly, was carried forward in December [2010] by a Human Rights Watch call on the United States "to suspend financing to Israel in an amount equivalent to the costs of Israel's spending in support of settlements," and to monitor contributions to Israel from tax-exempt U.S. organizations that violate international law, "including prohibitions against discrimination"—which would cast a wide net. Amnesty International had already called for an arms embargo on Israel.

The legitimation process also took a long step forward in December [2010], when Argentina, Bolivia and Brazil recognized the State of Palestine (Gaza and the West Bank), bringing the number of supporting nations to more than one hundred.

International lawyer John Whitbeck estimates that 80–90 percent of the world's population live in states that recognize Palestine, while 10–20 percent recognize the Republic of Kosovo.

The United States recognizes Kosovo but not Palestine. Accordingly, as Whitbeck writes in *Counterpunch*, media "act as though Kosovo's independence were an accomplished fact while Palestine's independence is only an aspiration which can never be realized without Israeli-American consent," reflecting the normal workings of power in the international arena.

Given the scale of Israeli settlement of the West Bank, it has been argued for more a decade that the international consensus on a two-state settlement is dead, or mistaken (though evidently most of the world does not agree). Therefore those concerned with Palestinian rights should call for Israeli takeover of the entire West Bank, followed by an anti-apartheid struggle of the South African variety that would lead to full citizenship for the Arab population there.

The argument assumes that Israel (and its U.S. protector) would agree to the takeover. It is far more likely that Israel will instead continue the programs leading to annexation of the parts of the West Bank that it is developing, roughly half the area, and take no responsibility for the rest, thus defending itself from the "demographic problem"—too many non-Jews in a Jewish state—and meanwhile severing besieged Gaza from the rest of Palestine.

One analogy between Israel and South Africa merits attention. As apartheid was implemented, South African nationalists recognized they were becoming international pariahs. In 1958 the foreign minister informed the U.S. ambassador that U.N. condemnations and other protests were of little concern as long as South Africa was supported by the global hegemon—the United States. He proved to be fairly accurate.

By the 1970s, the U.N. declared an arms embargo, soon followed by boycott campaigns and divestment. South Africa reacted in ways calculated to enrage international opinion. In a gesture of contempt for the U.N. and President Jimmy Carter—who failed to react so as not to disrupt worthless negotiations—South Africa launched a murderous raid on the Cassinga refugee camp in Angola just as the Carter-led "contact group" was to present a settlement for Namibia.

The similarity to Israel's behavior today is striking—for example, the attack on Gaza in December 2008–January 2009 and on the Gaza freedom flotilla in May 2010.

When President Reagan took office in 1981, he lent support to South Africa's domestic crimes and its murderous depredations in neighboring countries. The policies were justified in the framework of the war on terror that Reagan had declared on coming into office. In 1988, his administration designated Nelson Mandela's African National Congress one of the world's "more notorious terrorist groups" (Mandela himself was only removed from Washington's "terrorist list" in 2008). South Africa was defiant, and even triumphant, with its internal enemies crushed, and enjoying solid support from the one state that mattered in the global system.

Shortly after, U.S. policy shifted. U.S. and South African business interests very likely realized they would be better off by maintaining the class structure and ending the apartheid burden. And apartheid soon collapsed.

South Africa is not the only recent case where ending U.S. support for crimes has led to significant quick and significant change. Another is East Timor. The U.S.

(along with Britain and other allies) strongly supported Indonesia's 1975 invasion and the incredible slaughter that followed, and continued to do so well into 1999, when atrocities in that year alone surpassed anything attributed to Serbia prior to the NATO bombing, and what preceded of course was far worse. Indonesia remained defiant, declaring that it would never leave. In mid-September, under considerable domestic and international pressure, Clinton reversed course and announced quietly that the game was over. Indonesian forces quickly withdrew and an Australian-led peacekeeping force entered unopposed by military forces. Power matters.

Can such a transformative shift happen in Israel's case, clearing the way to a diplomatic settlement? There are many barriers, among them the very close military and intelligence ties between the United States and Israel since 1967, with earlier roots.

The most outspoken support for Israeli crimes comes from the business world. The U.S. high-tech industry is closely integrated with its Israeli counterpart. To cite just one example, the world's largest chip manufacturer, Intel, is establishing its most advanced production unit in Israel.

A U.S. cable released by WikiLeaks reveals that Rafael military industries in Haifa is one of the sites considered vital to U.S. interests due to its production of cluster bombs; Rafael had already moved some operations to the United States to gain better access to United States aid and markets. There is also a powerful Israel lobby, though of course dwarfed by the business and military lobbies.

Critical cultural facts apply, too. Christian Zionism long precedes Jewish Zionism, and is not restricted to the one-third of the U.S. population that believes the Bible is the literal truth. When British General Edmund Allenby conquered Jerusalem in 1917, the national press declared him to be Richard the Lionhearted, finally rescuing the Holy Land from the infidels.

Next, Jews must return to the homeland promised to them by the Lord. Articulating a common elite view, Harold Ickes, Franklin Roosevelt's secretary of the interior, described Jewish colonization of Palestine as an achievement "without comparison in the history of the human race." Similar views have had considerable resonance among U.S. elites.

There is also an instinctive sympathy for a settler-colonial society that is seen to be retracing the history of the United States itself, bringing civilization to the lands that the undeserving natives had misused—doctrines deeply rooted in centuries of imperialism, and particularly strong in what is sometimes called "the Anglosphere," Britain's offshoots, settler-colonial societies themselves and commonly the strongest supporters of Israeli violence and expansion.

To break the logjam it will be necessary to dismantle the reigning illusion that the United States is an "honest broker" desperately seeking to reconcile recalcitrant adversaries, and to recognize that serious negotiations would be between the U.S.-Israel and the rest of the world.

If U.S. power centers can be compelled by popular opinion to abandon decades-old rejectionism, many prospects that seem remote might become suddenly possible.

The Arab Word Is on Fire

FEBRUARY 1, 2011

"The Arab world is on fire," *al-Jazeera* reported on January 27 [2011], while throughout the region, Western allies "are quickly losing their influence."

The shock wave was set in motion by the dramatic uprising in Tunisia that drove out a Western-backed dictator, with reverberations especially in Egypt, where demonstrators overwhelmed a dictator's brutal police.

Observers compared the events to the toppling of Russian domains in 1989, but there are important differences.

Crucially, no Mikhail Gorbachev exists among the great powers that support the Arab dictators. Rather, Washington and its allies keep to the well-established principle that democracy is acceptable only insofar as it conforms to strategic and economic objectives: fine in enemy territory (up to a point), but not in our backyard, please, unless it is properly tamed.

One 1989 comparison has some validity: Romania, where Washington maintained its support for Nicolae Ceausescu, the most vicious of the East European dictators, until the allegiance became untenable. Then Washington hailed his overthrow while the past was erased.

That is a standard pattern: Ferdinand Marcos, Jean-Claude Duvalier, Chun Doo Hwan, Suharto and many other useful gangsters. It may be under way in the case of Hosni Mubarak, along with routine efforts to try to ensure that a successor regime will not veer far from the approved path.

The current hope appears to be Mubarak loyalist Gen. Omar Suleiman, just named Egypt's vice president. Suleiman, the longtime head of the intelligence services, is despised by the rebelling public almost as much as the dictator himself.

A common refrain among pundits is that fear of radical Islam requires (reluctant) opposition to democracy on pragmatic grounds. While not without some merit, the formulation is misleading. The general threat has always been independence. In the Arab world, the United States and its allies have regularly supported radical Islamists, sometimes to prevent the threat of secular nationalism.

A familiar example is Saudi Arabia, the ideological center of radical Islam (and of Islamic terror). Another in a long list is Zia ul-Haq, the most brutal of Pakistan's dictators and President Reagan's favorite, who carried out a program of radical Islamization (with Saudi funding).

"The traditional argument put forward in and out of the Arab world is that there is nothing wrong, everything is under control," says Marwan Muasher, former Jordanian official and now director of Middle East research for the Carnegie Endowment. "With this line of thinking, entrenched forces argue that opponents and outsiders calling for reform are exaggerating the conditions on the ground."

Therefore the public can be dismissed. The doctrine traces far back and generalizes worldwide, to U.S. home territory as well. In the event of unrest, tactical shifts may be necessary, but always with an eye to reasserting control.

The vibrant democracy movement in Tunisia was directed against "a police state, with little freedom of expression or association, and serious human rights problems," ruled by a dictator whose family was hated for their venality. This was the assessment by U.S. Ambassador Robert Godec in a July 2009 cable released by WikiLeaks.

Relying on such assessments, some observers hold that the WikiLeaks "documents should create a comforting feeling among the American public that officials aren't asleep at the switch"—indeed, the cables are so supportive of U.S. policies that it is almost as if Obama is leaking them himself (or so Jacob Heilbrunn writes in *The National Interest*).

"America should give Assange a medal," says a headline in the *Financial Times*. Chief foreign policy analyst Gideon Rachman writes that "America's foreign policy comes across as principled, intelligent and pragmatic . . . the public position taken by the U.S. on any given issue is usually the private position as well."

In this view, WikiLeaks undermines the "conspiracy theorists" who question the noble motives that Washington regularly proclaims.

Godec's cable supports these judgments—at least if we look no further. If we do, as foreign policy analyst Stephen Zunes reports in *Foreign Policy in Focus*, we find, with Godec's information in hand, that Washington provided $12 million in military aid to Tunisia. As it happens, Tunisia was one of only five foreign beneficiaries: Israel (routinely); the two Middle East dictatorships Egypt and Jordan; and Colombia, which has long had the worst human-rights record and the most U.S. military aid in the hemisphere.

Heilbrunn's Exhibit A is Arab support for U.S. policies targeting Iran, revealed by leaked cables. Rachman too seizes on this example, as did the media generally, hailing these encouraging revelations. The reactions illustrate how profound is the contempt for democracy in the educated culture.

Unmentioned is what the population thinks—easily discovered. According to polls released by the Brookings Institution in August [2010], some Arabs agree with Washington and Western commentators that Iran is a threat: 10 percent. In contrast, they regard the United States and Israel as the major threats (77 percent; 88 percent).

Arab opinion is so hostile to Washington's policies that a majority (57 percent) thinks regional security would be enhanced if Iran had nuclear weapons. Still, "there is nothing wrong, everything is under control" (as Marwan Muasher describes the prevailing fantasy). The dictators support us. Their subjects can be ignored—unless they break their chains, and then policy must be adjusted.

Other leaks also appear to lend support to the enthusiastic judgments about Washington's nobility. In July 2009, Hugo Llorens, U.S. ambassador to Honduras, informed Washington of an embassy investigation of "legal and constitutional issues surrounding the June 28 forced removal of President Manuel 'Mel' Zelaya."

The embassy concluded that "there is no doubt that the military, Supreme Court and National Congress conspired on June 28 [2010] in what constituted an illegal and unconstitutional coup against the Executive Branch." Very admirable, except that President Obama proceeded to break with almost all of Latin America and Europe by supporting the coup regime and dismissing subsequent atrocities.

Perhaps the most remarkable WikiLeaks revelations have to do with Pakistan, reviewed by foreign policy analyst Fred Branfman in Truthdig.

The cables reveal that the U.S. embassy is well aware that Washington's war in Afghanistan and Pakistan not only intensifies rampant anti-Americanism but also "risks destabilizing the Pakistani state" and even raises a threat of the ultimate nightmare: that nuclear weapons might fall into the hands of Islamic terrorists.

Again, the revelations "should create a comforting feeling . . . that officials are not asleep at the switch" (Heilbrunn)—while Washington marches stalwartly toward disaster.

The Cairo-Madison Connection

MARCH 1, 2011

On February 20 [2011], Kamal Abbas, Egyptian union leader and prominent figure in the January 25 movement, sent a message to the "workers of Wisconsin": "We stand with you as you stood with us."

Egyptian workers have long fought for fundamental rights denied by the U.S.-backed Hosni Mubarak regime. Kamal is right to invoke the solidarity that has long been the driving force of the labor movement worldwide, and to compare their struggles for labor rights and democracy.

The two are closely intertwined. Labor movements have been in the forefront of protecting democracy and human rights and expanding their domains, a primary reason why they are the bane of systems of power, both state and private.

The trajectories of labor struggles in Egypt and the United States may be crossing, but they are heading in opposite directions: toward gaining rights in Egypt, and defending rights under harsh attack in the United States.

The two cases merit a closer look.

The January 25 uprising was sparked by the tech-savvy young people of the April 6 movement, which arose in Egypt in spring 2008 in "solidarity with striking textile workers in Mahalla," labor analyst Nada Matta observes.

State violence crushed the strike and solidarity actions, but Mahalla was "a symbol of revolt and challenge to the regime," Matta adds. The strike became particularly threatening to the dictatorship when workers' demands extended beyond their local concerns to a minimum wage for all Egyptians.

Matta's observations are confirmed by Joel Beinin, the leading U.S. authority on Egyptian labor. Over many years of struggle, Beinin reports, workers have established bonds and can mobilize readily.

When the workers joined the January 25 movement, the impact was decisive, and the military command sent Mubarak on his way. That was a great victory for the Egyptian democracy movement, though many barriers remain, internal and external.

The external barriers are clear. The United States and its allies cannot easily tolerate functioning democracy in the Arab world.

For evidence, look to public opinion polls in Egypt and throughout the Middle East. By overwhelming majorities, the public regard the United States and Israel as the major threats, not Iran. Indeed, most think that the region would be better off if Iran had nuclear weapons. One of many reasons why Washington and its allies definitely do not want public opinion expressed in policy, one measure of functioning democracy.

We can anticipate that Washington will keep to its traditional policy, well confirmed by scholarship: Democracy is tolerable only insofar as it conforms to strategic-economic objectives. The United States' fabled "yearning for democracy" is reserved for ideologues and propaganda.

The course of democracy in the United States has been complex. After World War II the country enjoyed unprecedented growth, largely egalitarian and accompanied by legislation that benefited most people. The trend continued through the Richard Nixon years, which ended the liberal era. The popular activism of the 1960s meanwhile substantially expanded democratic participation.

The backlash against the democratizing impact of 1960s activism and Nixon's class treachery was not long in coming: a vast increase in lobbying to shape legislation, establishing of right-wing think tanks to capture the ideological spectrum, and many other measures.

The economy also shifted course sharply toward financialization and export of production. Inequality soared, primarily due to the skyrocketing wealth of the top 1 percent of the population—or even a smaller fraction, limited to mostly CEOs, hedge fund managers and the like.

For the majority, real incomes stagnated. Most resorted to increased working hours, debt and asset inflation. Then came the $8 trillion housing bubble, unnoticed by the Federal Reserve and almost all economists, who were enthralled by efficient market dogmas. When the bubble burst, the economy collapsed to near Depression levels for manufacturing workers and many others.

Concentration of income confers political power, which in turn leads to legislation that further enhances the privilege of the super-rich: tax policies, deregulation, rules of corporate governance and much else.

Alongside this vicious cycle, costs of campaigning sharply increased, driving both political parties to cater to the corporate sector—the Republicans reflexively, and the Democrats (now pretty much the moderate Republicans of earlier years) following not far behind.

In 1978, as the process was taking off, United Auto Workers President Doug Fraser condemned business leaders for having "chosen to wage a one-sided class war in this country—a war against working people, the unemployed, the poor, the minorities, the very young and the very old, and even many in the middle class of our society," and having "broken and discarded the fragile, unwritten compact previously existing during a period of growth and progress." Correct, though the realization was much too late.

As working people won basic rights in the 1930s, business leaders warned of "the hazard facing industrialists in the rising political power of the masses" and called for urgent measures to beat back the threat, as discussed by Alex Carey in *Taking the Risk Out of Democracy*, his pioneering study of corporate propaganda. They understood as well as Mubarak that unions are a leading force in advancing rights and democracy. In the United States, unions are the primary counterforce to corporate tyranny.

By now, U.S. private-sector unions have been severely weakened under intensive state-corporate assault. Public-sector unions have recently come under sharp attack from right-wing opponents who cynically exploit the economic crisis caused primarily by the finance industry and its associates in government.

Popular anger must be diverted from the agents of the financial crisis, who are profiting from it; for example, Goldman Sachs, "on track to pay out $17.5 billion in compensation for last year," the business press reports, with CEO Lloyd Blankfein receiving a $12.6 million bonus while his base salary more than triples to $2 million.

Instead, propaganda must blame teachers and other public-sector workers with their fat salaries and exorbitant pensions—all a fabrication, on a model that is all too familiar. For Wisconsin's Governor Scott Walker, along with other Republicans and many Democrats, the slogan is that austerity must be shared—with some notable exceptions.

The propaganda has been fairly effective. Walker can count on at least a large minority to support his brazen effort to destroy the unions. Invoking the deficit as an excuse is pure farce.

In different ways, the fate of democracy is at stake in Madison, Wisconsin, no less than it is in Tahrir Square.

Libya and the World of Oil

APRIL 1, 2011

Last month [March 2011], at the international tribunal on crimes during the civil war in Sierra Leone, the prosecution rested its case in the trial of former Liberian president Charles Taylor.

The chief prosecutor, U.S. law professor David Crane, informed *The Times of London* that the case was incomplete: The prosecutors intended to charge Moammar Gadhafi, who, Crane said, "was ultimately responsible for the mutilation, maiming and/or murder of 1.2 million people."

But the charge was not to be. The United States, Britain and others intervened to block it. Asked why, Crane said, "Welcome to the world of oil."

Another recent Gadhafi casualty was Sir Howard Davies, the director of the London School of Economics, who resigned after revelations of the school's links to the Libyan dictator.

In Cambridge, Massachusetts, the Monitor Group, a consultancy firm founded by Harvard Business School professors, was well paid for such services as a book to bring Gadhafi's immortal words to the public "in conversation with renowned international experts" who were brought to Libya to converse with the great man, along with other efforts "to enhance international appreciation of [Gadhafi's] Libya."

The world of oil is rarely far in the background in affairs

concerning this region. It provides useful guidelines for Western reactions to the remarkable democracy uprisings in the Arab world. An oil-rich dictator who is a reliable client is granted virtual free rein. There was little reaction when Saudi Arabia declared on March 5 [2011], "Laws and regulations in the Kingdom totally prohibit all kinds of demonstrations, marches and sit-in protests [and prohibits] calling for them as they go against the principles of Shariah and Saudi customs and traditions." The kingdom mobilized huge security forces that rigorously enforced the ban.

In Kuwait, small demonstrations were crushed. The mailed fist struck in Bahrain after Saudi-led military forces intervened to ensure that the minority Sunni monarchy would not be threatened by calls for democratic reforms by the Shiite majority and others.

Bahrain is sensitive not only because it hosts the U.S. Fifth Fleet but also because it is across the causeway from predominantly Shiite areas of Saudi Arabia, the location of most of the kingdom's oil. The world's primary energy resources happen to be located near the northern Persian Gulf (or Arabian Gulf, as Arabs often call it), largely Shiite, a potential nightmare for Western planners.

In Egypt and Tunisia, the popular uprising has won impressive victories, but as the Carnegie Endowment reported, the regimes remain and are "seemingly determined to curb the pro-democracy momentum generated so far. A change in ruling elites and system of governance is still a distant goal"—and one that the West will seek to keep far removed.

Libya is a different case, an oil-rich state run by a brutal dictator who, however, is unreliable: A dependable client would be far preferable. When nonviolent protests erupted, Gadhafi moved quickly to crush them.

On March 22 [2011], as Gadhafi's forces were converging on the rebel capital of Benghazi, Obama's top Middle East adviser Dennis Ross warned that if there is a massacre, "everyone would blame us for it," an unacceptable consequence.

And the West didn't want Gadhafi to enhance his power and independence by crushing the rebellion. The United States joined in the U.N. Security Council authorization of a "no-fly zone" and measures to protect civilians, to be implemented primarily by France, the United Kingdom and the United States, the three traditional imperial powers (though we may recall—Libyans presumably do— that after World War I, Italy conducted virtual genocide in eastern Libya).

The no-fly zone prevented a likely massacre, but the U.N. resolution was immediately rejected in practice by the triumvirate and interpreted as authorizing direct participation in support for the rebels. A cease-fire was imposed on Gadhafi's forces, but the rebels were helped to advance to the West. In short order they conquered the major sources of Libya's oil production, at least temporarily.

On March 28 [2011], the London-based Arab journal *Al-Quds Al-Arabi* warned that the intervention may leave Libya with "two states, a rebel-held, oil-rich East and a poverty-stricken, Gadhafi-led West. . . . Given that the oil wells have been secured, we may find ourselves facing a new Libyan oil emirate, sparsely inhabited, protected by the West and very similar to the Gulf's emirate states." Or the Western-backed rebellion might proceed all the way to eliminate the irritating dictator.

It is commonly argued that oil cannot be a motive for the intervention because the West had access to the prize under Gadhafi. True but irrelevant. The same could be said about Iraq under Saddam Hussein, or Iran and Cuba today.

What the West seeks is what Bush announced with regard to Iraq in 2007–8, when the goals of the invasion could no longer be concealed: control, or at least dependable clients, and in the case of Libya, access to vast unexplored areas expected to be rich in oil.

The Western powers are acting in virtual isolation. States in the region—Turkey and Egypt—kept away, while Africa called for diplomacy and negotiation, as did the BRICS countries (Brazil, Russia, India, China, South Africa), meeting in China. The Gulf dictators would be happy to see Gadhafi gone—but, even as they're groaning under the weight of advanced weapons provided to them to recycle petrodollars and ensure obedience, they barely offer more than token participation. The same is true beyond.

The Arab Spring has deep roots. The region has been simmering for years. The first of the current wave of protests began last year in Western Sahara, the last African colony, invaded by Morocco in 1975 and illegally held since, in a manner similar to East Timor and the Israeli-occupied territories.

A nonviolent protest last November [2010] was crushed by Moroccan forces. France intervened to block a Security Council inquiry into the crimes of its client.

Then a flame ignited in Tunisia that has since spread into a conflagration.

The International Assault on Labor

MAY 2, 2011

In most of the world, May Day is an international workers' holiday, bound up with the bitter nineteenth-century struggle of American workers for an eight-hour day. The May Day just past leads to somber reflection.

A decade ago, a useful word was coined in honor of May Day by radical Italian labor activists: "precarity." It referred at first to the increasingly precarious existence of working people "at the margins"—women, youth, migrants. Then it expanded to apply to the growing "precariat" of the core labor force, the "precarious proletariat" suffering from the programs of deunionization, flexibilization and deregulation that are part of the assault on labor throughout the world.

By that time, even in Europe there was mounting concern about what labor historian Ronaldo Munck, citing Ulrich Beck, calls the "Brazilianization of the West . . . the spread of temporary and insecure employment, discontinuity and loose informality into Western societies that have hitherto been the bastions of full employment." By now, an insult to Brazil, which is seeking to address such problems, not exacerbate them.

The state-corporate war against unions has recently extended to the public sector, with legislation to ban collective bargaining and other elementary rights. Even in pro-labor Massachusetts, the House of Representatives voted right before May Day to sharply restrict the rights of police officers, teachers and other municipal employees to bargain over health care—essential matters in the United States, with its dysfunctional and highly inefficient privatized health care system.

The rest of the world may associate May 1 with the struggle of American workers for basic rights, but in the United States that solidarity is suppressed in favor of a jingoist holiday. May 1 is "Loyalty Day," designated by Congress in 1958 for "the reaffirmation of loyalty to the United States and for the recognition of the heritage of American freedom."

President Eisenhower proclaimed further that Loyalty Day is also Law Day, reaffirmed annually by displaying the flag and dedication to "Justice for All," "Foundations of Freedom" and "Struggle for Justice."

The U.S. calendar has a Labor Day, in September, celebrating the return to work after a vacation that is far briefer than in other industrial countries.

The ferocity of the assault against labor by the U.S. business class is illustrated by Washington's failure, for sixty years, to ratify the core principle of international labor law which guarantees freedom of association. Legal analyst Steve Charnovitz calls it "the untouchable treaty in American politics" and observes that there has never even been any debate about the matter.

Washington's dismissal of some conventions supported by the International Labor Organization contrasts sharply with its dedication to enforcement of monopoly-pricing rights for corporations, disguised under the mantle of "free trade" in one of the contemporary Orwellisms.

In 2004, the ILO reported that "economic and social

insecurities were multiplying with globalization and the policies associated with it, as the global economic system has become more volatile and workers were increasingly shouldering the burden of risk, for instance, though pension and health care reforms."

This was what economists call the period of the Great Moderation, hailed as "one of the great transformations of modern history," led by the United States and based on "liberation of markets," particularly "deregulation of financial markets."

This paean to the American way of free markets was delivered by *Wall Street Journal* editor Gerard Baker in January 2007, just months before the system crashed—and with it the entire edifice of the economic theology on which it was based—bringing the world economy to near disaster.

The crash left the United States with levels of real unemployment comparable to the Great Depression, and in many ways worse, because under the current policies of the masters those jobs are not coming back, as they did through massive government stimulus during World War II and the following decades of the "golden age" of state capitalism.

During the Great Moderation, American workers had become accustomed to a precarious existence. The rise of an American precariat was proudly hailed as a primary factor in the Great Moderation that brought slower economic growth, virtual stagnation of real income for the majority of the population, and wealth beyond the dreams of avarice for a tiny sector, mostly the agents of this historical transformation.

The high priest of this magnificent economy was Alan

Greenspan, described by the business press as "saintly" for his brilliant stewardship. Glorying in his achievements, he testified before Congress that they relied in part on "atypical restraint on compensation increases [which] appears to be mainly the consequence of greater worker insecurity."

The disaster of the Great Moderation was salvaged by heroic government efforts to reward the perpetrators. Neil Barofsky, stepping down on March 30 [2011] as special inspector general of the bailout program, wrote a revelatory *New York Times* op-ed about how the bailout worked.

In theory, the legislative act that authorized the bailout was a bargain: The financial institutions would be saved by the taxpayer, and the victims of their misdeeds would be somewhat compensated by measures to protect home values and preserve homeownership.

Part of the bargain was kept: The financial institutions were rewarded lavishly for causing the crisis, and forgiven for outright crimes. But the rest of the program floundered.

As Barofsky writes: "Foreclosures continue to mount, with 8 million to 13 million filings forecast over the program's lifetime," while "the biggest banks are 20 percent larger than they were before the crisis and control a larger part of our economy than ever. They reasonably assume that the government will rescue them again, if necessary. Indeed, credit rating agencies incorporate future government bailouts into their assessments of the largest banks, exaggerating market distortions that provide them with an unfair advantage over smaller institutions, which continue to struggle."

In short, President Obama's programs were "a giveaway to Wall Street executives" and a blow in the solar plexus to their defenseless victims.

The outcome should surprise only those who insist on hopeless naïveté about the design and implementation of policy, particularly when economic power is highly concentrated and state capitalism has entered into a new stage of "creative destruction," to borrow Joseph Schumpeter's famous phrase, but with a twist: creative in ways to enrich and empower the rich and powerful, while the rest are free to survive as they may, while celebrating Loyalty and Law Day.

The Revenge Killing
of Osama Bin Laden

JUNE 1, 2011

The May 1 [2011] U.S. attack on Osama bin Laden's compound violated multiple elementary norms of international law, beginning with the invasion of Pakistani territory.

There appears to have been no attempt to apprehend the unarmed victim, as presumably could have been done by the seventy-nine commandos facing almost no opposition.

President Obama announced that "justice has been done." Many did not agree—even close allies.

British barrister Geoffrey Robertson, who generally supported the operation, nevertheless described Obama's claim as an "absurdity" that should have been obvious to a former professor of constitutional law.

Pakistani and international law require inquiry "whenever violent death occurs from government or police action," Robertson points out. Obama undercut that possibility with a "hasty 'burial at sea' without a post mortem, as the law requires."

"It was not always thus," Robertson usefully reminds us: "When the time came to consider the fate of men much more steeped in wickedness than Osama bin Laden—namely the Nazi leadership—the British government wanted them hanged within six hours of capture.

"President Truman demurred, citing the conclusion of

Justice Robert Jackson (chief prosecutor at the Nuremberg trial) that summary execution 'would not sit easily on the American conscience or be remembered by our children with pride. . . . The only course is to determine the innocence or guilt of the accused after a hearing as dispassionate as the times will permit and upon a record that will leave our reasons and motives clear.'"

Another perspective on the attack comes in a report in *The Atlantic* by veteran Middle East and military correspondent Yochi Dreazen and colleagues. Citing a "senior U.S. official," they conclude that the bin Laden killing was a planned assassination.

"For many at the Pentagon and the Central Intelligence Agency who had spent nearly a decade hunting bin Laden, killing the militant was a necessary and justified act of vengeance," they write.

Furthermore, "capturing bin Laden alive would have also presented the administration with an array of nettlesome legal and political challenges."

They quote former West German Chancellor Helmut Schmidt, who commented that "the U.S. raid was 'quite clearly a violation of international law' and that bin Laden should have been detained and put on trial."

They contrast Schmidt with U.S. Attorney General Eric Holder, who "defended the decision to kill bin Laden although he didn't pose an immediate threat to the Navy SEALs," and testified to Congress that the assault had been "lawful, legitimate and appropriate in every way."

They observe further that the assassination is "the clearest illustration to date" of a crucial distinction between the Bush and Obama counterterror policies. Bush captured suspects and sent them to Guantánamo and other camps, with consequences now well known. Obama's policy is to kill suspects (along with "collateral damage").

The roots of the revenge killing are deep. In the immediate aftermath of 9/11, the American desire for vengeance displaced concern for law or security.

In his book *The Far Enemy*, Fawaz Gerges, a leading academic specialist on the jihadi movement, found that "the dominant response by jihadis to September 11 is an explicit rejection of al-Qaida and total opposition to the internationalization of jihad. . . . Al-Qaida united all social forces [in the Muslim world] against its global jihad."

The influential Lebanese cleric Sheikh Mohammed Hussein Fadlallah sharply condemned al-Qaida's 9/11 atrocities on principled grounds. "We must not punish individuals who have no relationship with the American administration or even those who have an indirect role," he said.

Fadlallah was the target of a CIA-organized assassination operation in 1985, a huge truck bomb placed outside a mosque. He escaped, but eighty others were killed, mostly women and girls, as they left the mosque—one of those innumerable crimes that don't enter the annals of terror.

Subsequent U.S. actions, particularly the invasion of Iraq, gave new life to al-Qaida.

What are the likely consequences of the killing of bin Laden? For the Arab world, it will probably mean little. He had long been a fading presence, and in the past few months was eclipsed by the Arab Spring.

A fairly general perception in the Arab world is captured by the headline in a Lebanese newspaper: "The execution of bin Laden: A settling of accounts between killers."

The most immediate and significant consequences are likely to be seen in Pakistan. There is much discussion of Washington's anger that Pakistan didn't turn over bin Laden. Less is said about the fury in Pakistan that the United States invaded its territory to carry out a political assassination.

Pakistan is the most dangerous country on Earth, with the fastest-growing nuclear arsenal. The revenge killing on Pakistani soil only stoked the anti-American fervor that had long been building.

In his new book, *Pakistan: A Hard Country*, Anatol Lieven writes that "if the U.S. ever put Pakistani soldiers in a position where they felt that honor and patriotism required them to fight America, many would be very glad to do so."

And if Pakistan collapsed, an "absolutely inevitable result would be the flow of large numbers of highly trained ex-soldiers, including explosive experts and engineers, to extremist groups."

The primary threat is leakage of fissile materials to jihadi hands, a horrendous eventuality.

The Pakistani military has already been pushed to the edge by U.S. attacks on Pakistani sovereignty. One factor is the drone attacks in Pakistan that Obama escalated immediately after the killing of bin Laden, rubbing salt in the wounds.

But there is much more, including the demand that the Pakistani military cooperate in the U.S. war against the Afghan Taliban. The overwhelming majority of Pakistanis see the Taliban as fighting a just war of resistance against an invading army, according to Lieven.

The killing of bin Laden could have been the spark

that set off a conflagration, with dire consequences, particularly if the invading force had been compelled to fight its way out, as was anticipated.

Perhaps the assassination was perceived as an "act of vengeance," as Robertson concludes. Whatever the motive, it could hardly have been security.

In Israel, a Tsunami Warning

JULY 6, 2011

In May [2011], in a closed meeting of many of Israel's business leaders, Idan Ofer, a holding-company magnate, warned, "We are quickly turning into South Africa. The economic blow of sanctions will be felt by every family in Israel."

The business leaders' particular concern was the U.N. General Assembly session this September [2011], where the Palestinian Authority is planning to call for recognition of a Palestinian state.

Dan Gillerman, Israel's former ambassador to the United Nations, warned participants that "the morning after the anticipated announcement of recognition of a Palestinian state, a painful and dramatic process of South-africanization will begin"—meaning that Israel would become a pariah state, subject to international sanctions.

In this and subsequent meetings, the oligarchs urged the government to initiate efforts modeled on the Saudi (Arab League) proposals and the unofficial Geneva Accord of 2003, in which high-level Palestinian and Israeli negotiators detailed a two-state settlement that was welcomed by most of the world, dismissed by Israel and ignored by Washington.

In March [2011], Israel's defense minister, Ehud Barak, warned of the prospective U.N. action as a "tsunami." The fear is that the world will condemn Israel not only for violating international law but also for carrying out its criminal acts in an occupied state recognized by the United Nations.

The United States and Israel are waging intensive diplomatic campaigns to head off the tsunami. If they fail, recognition of a Palestinian state is likely.

More than one hundred states already recognize Palestine. The United Kingdom, France and other European nations have upgraded the Palestine General Delegation to "diplomatic missions and embassies—a status normally reserved only for states," Victor Kattan observes in the *American Journal of International Law*.

Palestine has also been admitted to U.N. organizations apart from UNESCO and the World Health Organization, which have avoided the issue for fear of U.S. defunding—no idle threat.

In June [2011] the U.S. Senate passed a resolution threatening to suspend aid for the Palestine Authority if it persists with its U.N. initiative. Susan Rice, U.S. ambassador to the United Nations, warned that there was "no greater threat" to U.S. funding of the United Nations "than the prospect of Palestinian statehood being endorsed by member states," London's *Daily Telegraph* reports. Israel's new U.N. Ambassador, Ron Prosor, informed the Israeli press that U.N. recognition "would lead to violence and war."

The United Nations would presumably recognize Palestine in the internationally accepted borders, including the West Bank and Gaza, with the Golan Heights returned to Syria. The heights were annexed by Israel in December 1981, in violation of U.N. Security Council orders.

In the West Bank, the settlements and acts to support them are clearly in violation of international law, as affirmed by the World Court and the Security Council.

In February 2006, the United States and Israel imposed a siege in Gaza after the "wrong side"—Hamas—won elections in Palestine, recognized as free and fair. The siege became much harsher in June 2007 after the failure of a U.S.-backed military coup to overthrow the elected government.

In June 2010, the siege of Gaza was condemned by the International Committee of the Red Cross—which rarely issues such reports—as "collective punishment imposed in clear violation" of international humanitarian law. The BBC reported that the ICRC "paints a bleak picture of conditions in Gaza: hospitals short of equipment, power cuts lasting hours each day, drinking water unfit for consumption," and the population of course imprisoned.

The criminal siege extends the U.S.-Israeli policy since 1991 of separating Gaza from the West Bank, thus ensuring that any eventual Palestinian state would be effectively contained within hostile powers—Israel and the Jordanian dictatorship. The Oslo Accords, signed by Israel and the Palestine Liberation Organization in 1993, proscribe separating Gaza from the West Bank.

A more immediate threat facing U.S.-Israeli rejectionism is the Freedom Flotilla that seeks to challenge the blockade of Gaza by bringing letters and humanitarian aid. In May 2010, the last such attempt led to an attack by Israeli commandos in international waters—a major crime in itself—in which nine passengers were killed, actions bitterly condemned outside the United States.

In Israel, most people convinced themselves that the commandos were the innocent victims, attacked by passengers, another sign of the self-destructive irrationality sweeping the society.

Today the United States and Israel are vigorously seeking to block the flotilla. U.S. Secretary of State Hillary Clinton virtually authorized violence, stating that "Israelis have the right to defend themselves" if flotillas "try to provoke action by entering into Israeli waters"—that is, the territorial waters of Gaza, as if Gaza belonged to Israel.

Greece agreed to prevent the boats from leaving (that is, those boats not already sabotaged)—though, unlike Clinton, Greece referred rightly to "the maritime area of Gaza."

In January 2009, Greece had distinguished itself by refusing to permit U.S. arms to be shipped to Israel from Greek ports during the vicious U.S.-Israeli assault in Gaza. No longer an independent country in its current financial duress, Greece evidently cannot risk such unusual integrity.

Asked whether the flotilla is a "provocation," Chris Gunness, the spokesperson for the U.N. Relief and Works Agency, the major aid agency for Gaza, described the situation as desperate: "If there were no humanitarian crisis, if there weren't a crisis in almost every aspect of life in Gaza there would be no need for the flotilla. . . . 95 percent of all water in Gaza is undrinkable, 40 percent of all disease is water-borne . . . 45.2 percent of the labor force is unemployed, 80 percent aid dependency, a tripling of the abject poor since the start of the blockade. Let's get rid of this blockade and there would be no need for a flotilla."

Diplomatic initiatives such as the strategy for a Palestinian state, and nonviolent actions generally, threaten those who hold a virtual monopoly on violence. The United States and Israel are trying to sustain indefensible positions: the occupation and its subversion of the overwhelming, long-standing consensus on a diplomatic settlement.

America in Decline

AUGUST 5, 2011

"It is a common theme" that the United States, which "only a few years ago was hailed to stride the world as a colossus with unparalleled power and unmatched appeal—is in decline, ominously facing the prospect of its final decay," Giacomo Chiozza writes in the current *Political Science Quarterly*.

The theme is indeed widely believed. And with some reason, though a number of qualifications are in order. To start with, the decline has proceeded since the high point of U.S. power after World War II, and the remarkable triumphalism of the post–Gulf War 1990s was mostly self-delusion.

Another common theme, at least among those who are not willfully blind, is that American decline is in no small measure self-inflicted. The comic opera in Washington this summer [2011], which disgusts the country and bewilders the world, may have no analogue in the annals of parliamentary democracy.

The spectacle is even coming to frighten the sponsors of the charade. Corporate power is now concerned that the extremists they helped put in office may in fact bring down the edifice on which their own wealth and privilege relies, the powerful nanny state that caters to their interests.

Corporate power's ascendancy over politics and society—by now mostly financial—has reached the point that both political organizations, which at this stage barely resemble traditional parties, are far to the right of the population on the major issues under debate.

For the public, the primary domestic concern is unemployment. Under current circumstances, that crisis can be overcome only by a significant government stimulus, well beyond the recent one, which barely matched decline in state and local spending—though even that limited initiative probably saved millions of jobs.

For financial institutions the primary concern is the deficit. Therefore, only the deficit is under discussion. A large majority of the population favor addressing the deficit by taxing the very rich (72 percent, 27 percent opposed), reports a *Washington Post–ABC News* poll. Cutting health programs is opposed by overwhelming majorities (69 percent Medicaid, 78 percent Medicare). The likely outcome is therefore the opposite.

The Program on International Policy Attitudes surveyed how the public would eliminate the deficit. PIPA director Steven Kull writes, "Clearly both the administration and the Republican-led House [of Representatives] are out of step with the public's values and priorities in regard to the budget."

The survey illustrates the deep divide: "The biggest difference in spending is that the public favored deep cuts in defense spending, while the administration and the House propose modest increases. . . . The public also favored more spending on job training, education and pollution control than did either the administration or the House."

The final "compromise"—more accurately, capitulation to the far right—is the opposite throughout, and is almost certain to lead to slower growth and long-term harm to all but the rich and the corporations, which are enjoying record profits.

Not even discussed is that the deficit would be eliminated if, as economist Dean Baker has shown, the dysfunctional privatized health care system in the United States were replaced by one similar to that of other industrial societies, which have half the per-capita costs and health outcomes that are comparable or better.

The financial institutions and Big Pharma are far too powerful for such that of options even to be considered, though the thought seems hardly Utopian. Off the agenda for similar reasons are other economically sensible options, such as a small financial transactions tax.

Meanwhile new gifts are regularly lavished on Wall Street. The House Appropriations Committee cut the budget request for the Securities and Exchange Commission, the prime barrier against financial fraud, though by now so corrupted by its intimate relations with those allegedly regulated that a case can be made for its elimination. The Consumer Protection Agency is unlikely to survive intact.

Congress wields other weapons in its battle against future generations. Faced with Republican opposition to environmental protection, American Electric Power, a major utility, shelved "the nation's most prominent effort to capture carbon dioxide from an existing coal-burning power plant, dealing a severe blow to efforts to rein in emissions responsible for global warming," the *New York Times* reported.

The self-inflicted blows, while increasingly powerful, are not a recent innovation. They trace back to the 1970s, when the national political economy underwent major transformations, ending what is commonly called "the Golden Age" of (state) capitalism.

Two major elements were financialization (the shift of investor preference from industrial production to so-called FIRE: finance, insurance, real estate) and the offshoring of production. The ideological triumph of "free market doctrines," highly selective as always, administered further blows, as they were translated into tax and other fiscal policies, deregulation, rules of corporate governance linking huge CEO rewards to short-term profit, and other such policy decisions.

The resulting concentration of wealth yielded greater political power, accelerating a vicious cycle that has led to extraordinary wealth for a fraction of 1 percent of the population, mainly, while for the large majority real incomes have virtually stagnated.

In parallel, the cost of elections skyrocketed, driving both parties even deeper into corporate pockets. What remains of political democracy has been undermined further as both parties have turned to auctioning congressional leadership positions, as political economist Thomas Ferguson outlines in the *Financial Times*.

"The major political parties borrowed a practice from big box retailers like Walmart, Best Buy or Target," Ferguson writes. "Uniquely among legislatures in the developed world, U.S. congressional parties now post prices for key slots in the lawmaking process." The legislators who contribute the most funds to the party get the posts.

The result, according to Ferguson, is that debates "rely heavily on the endless repetition of a handful of slogans that have been battle-tested for their appeal to national investor blocs and interest groups that the leadership relies on for resources." The country be damned.

Before the 2007 crash for which they were largely responsible, the new post–Golden Age financial institutions had gained startling economic power, more than tripling their share of corporate profits. After the crash, a number of economists began to inquire into their function in purely economic terms. Nobel laureate Robert Solow concludes that their general impact may be negative: "The successes probably add little or nothing to the efficiency of the real economy, while the disasters transfer wealth from taxpayers to financiers."

By shredding the remnants of political democracy, the financial institutions lay the basis for carrying the lethal process forward—as long as their victims are willing to suffer in silence.

After 9/11, Was War the Only Option?

SEPTEMBER 5, 2011

This is the tenth anniversary of the horrendous atrocities of September 11, 2001, which, it is commonly held, changed the world.

The impact of the attacks is not in doubt. Just keeping to western and central Asia: Afghanistan is barely surviving, Iraq has been devastated and Pakistan is edging closer to a disaster that could be catastrophic.

On May 1, 2011, the presumed mastermind of the crime, Osama bin Laden, was assassinated in Pakistan. The most immediate significant consequences have also occurred in Pakistan. There has been much discussion of Washington's anger that Pakistan didn't turn over bin Laden. Less has been said about the fury among Pakistanis that the United States invaded their territory to carry out a political assassination. Anti-American fervor had already intensified in Pakistan, and these events have stoked it further.

One of the leading specialists on Pakistan, British military historian Anatol Lieven, wrote in *The National Interest* in February [2011] that the war in Afghanistan is "destabilizing and radicalizing Pakistan, risking a geopolitical catastrophe for the United States—and the world—which would dwarf anything that could possibly occur in Afghanistan."

At every level of society, Lieven writes, Pakistanis

overwhelmingly sympathize with the Afghan Taliban, not because they like them but because "the Taliban are seen as a legitimate force of resistance against an alien occupation of the country," much as the Afghan mujahedeen were perceived when they resisted the Russian occupation in the 1980s.

These feelings are shared by Pakistan's military leaders, who bitterly resent U.S. pressures to sacrifice themselves in Washington's war against the Taliban. Further bitterness comes from the terror attacks (drone warfare) by the United States within Pakistan, the frequency of which was sharply accelerated by President Obama; and from U.S. demands that the Pakistani army carry Washington's war into tribal areas of Pakistan that had been pretty much left on their own, even under British rule.

The military is the stable institution in Pakistan, holding the country together. U.S. actions might "provoke a mutiny of parts of the military," Lieven writes, in which case "the Pakistani state would collapse very quickly indeed, with all the disasters that this would entail."

The potential disasters are drastically heightened by Pakistan's huge, rapidly growing nuclear weapons arsenal, and by the country's substantial jihadi movement.

Both of these are legacies of the Reagan administration. Reagan officials pretended they did not know that Zia ul-Haq, the most vicious of Pakistan's military dictators and a Washington favorite, was developing nuclear weapons and carrying out a program of radical Islamization of Pakistan with Saudi funding.

The catastrophe lurking in the background is that these two legacies might combine, with fissile materials leaking into the hands of jihadis. Thus we might see nuclear weapons, most likely "dirty bombs," exploding in London and New York.

Lieven summarizes: "U.S. and British soldiers are in effect dying in Afghanistan in order to make the world more dangerous for American and British peoples."

Surely Washington understands that U.S. operations in what has been christened "Afpak"—Afghanistan-Pakistan—might destabilize and radicalize Pakistan.

The most significant WikiLeaks documents to have been released so far are the cables from U.S. Ambassador Anne Patterson in Islamabad, who supports U.S. actions in Afpak but warns that they "risk destabilizing the Pakistani state, alienating both the civilian government and military leadership, and provoking a broader governance crisis in Pakistan."

Patterson writes of the possibility that "someone working in [Pakistani government] facilities could gradually smuggle enough fissile material out to eventually make a weapon," a danger enhanced by "the vulnerability of weapons in transit."

A number of analysts have observed that bin Laden won some major successes in his war against the United States.

As Eric S. Margolis writes in *The American Conservative* in May [2011], bin Laden "repeatedly asserted that the only way to drive the U.S. from the Muslim world and defeat its satraps was by drawing Americans into a series of small but expensive wars that would ultimately bankrupt them."

That Washington seemed bent on fulfilling bin Laden's wishes was evident immediately after the 9/11 attacks.

In his 2004 book *Imperial Hubris*, Michael Scheuer,

a senior CIA analyst who had tracked Osama bin Laden since 1996, explains: "Bin Laden has been precise in telling America the reasons he is waging war on us. [He] is out to drastically alter U.S. and Western policies toward the Islamic world," and largely achieved his goal.

He continues: "U.S. forces and policies are completing the radicalization of the Islamic world, something Osama bin Laden has been trying to do with substantial but incomplete success since the early 1990s. As a result, I think it is fair to conclude that the United States of America remains bin Laden's only indispensable ally." And arguably remains so, even after his death.

The succession of horrors across the past decade leads to the question: *Was there an alternative to the West's response to the 9/11 attacks?*

The jihadi movement, much of it highly critical of bin Laden, could have been split and undermined after 9/11, if the "crime against humanity," as the attacks were rightly called, had been approached as a crime, with an international operation to apprehend the suspects. That was recognized at the time, but no such idea was even considered in the rush to war. It is worth adding that bin Laden was condemned in much of the Arab world for his part in the attacks.

By the time of his death, bin Laden had long been a fading presence, and in the previous months was eclipsed by the Arab Spring. His significance in the Arab world is captured by the headline in a *New York Times* article by Middle East specialist Gilles Kepel: "Bin Laden Was Dead Already."

That headline might have been dated far earlier, had the United States not mobilized the jihadi movement with retaliatory attacks on Afghanistan and Iraq.

Within the jihadi movement, bin Laden was doubtless a venerated symbol but apparently didn't play much more of a role for al-Qaida, this "network of networks," as analysts call it, which undertake mostly independent operations.

Even the most obvious and elementary facts about the decade lead to bleak reflections when we consider 9/11 and its consequences, and what they portend for the future.

NOTE

1. Adapted from *9-11: Was There an Alternative?*, the 10th-anniversary edition of *9-11*, by Noam Chomsky, published in the Open Media Series by Seven Stories Press. Reprinted with permission.

The Threat of Warships on an "Island of World Peace"

OCTOBER 5, 2011

Jeju Island, fifty miles southeast of South Korea's mainland, has been called the most idyllic place on the planet. The pristine, 706-square-mile volcanic island comprises three UNESCO World Natural Heritage sites.

Jeju's history, however, is far from idyllic. In 1948, two years before the outbreak of the Korean War, the islanders staged an uprising to protest, among other issues, the division of the Korean Peninsula into North and South. The mainland government, then under U.S. military occupation, cracked down on the Jeju insurgents.

South Korean police and military forces massacred islanders and destroyed villages. Korea historian John Merrill estimates that the death toll may have exceeded thirty thousand, about 15 percent of the island's population.

Decades later, a government commission investigated the Jeju uprising. In 2005, Roh Moo-hyun, then South Korea's president, apologized for the atrocities and designated Jeju an "Island of World Peace."

Today Jeju Island is once again threatened by joint U.S.-South Korean militarization and violence: the construction of a naval base on what many consider to be Jeju's most beautiful coastline.

For more than four years, island residents and peace

activists have engaged in determined resistance to the base, risking their lives and freedom.

The stakes are high for the world as well. Recently the Korean *JoongAng Daily*, in Seoul, described the island as "the spearhead of the country's defense line"—a line recklessly located three hundred miles from China.

In these troubled waters, the Jeju base would host up to twenty American and South Korean warships, including submarines, aircraft carriers and destroyers, several of which would be fitted with the Aegis ballistic-missile defense system.

For the United States, the base's purpose is to project force toward China—and to provide a forward operating installation in the event of a military conflict. The last thing the world needs is brinksmanship between the United States and China.

The protest now taking place on Jeju counts as a critical struggle against a potentially devastating war in Asia, and against the deeply rooted institutional structures that are driving the world toward ever more conflict.

Not surprisingly, China sees the base as a threat to its national security. At the very least, the base is likely to trigger confrontation and an arms race between South Korea and China, with the United States almost inevitably involved. Failure to prevent this dangerous, destructive project may well have consequences reaching far beyond Asia.

We need not speculate how Washington would react were China to establish a base near the U.S. coast.

The new base on Jeju is located in Gangjeong, a farming and fishing village that has reluctantly become the site of an epic battle for peace.

The resistance is a grassroots movement that goes well beyond the issue of the island's militarization. Human rights, the environment and free speech are also at stake. Though small and remote, Gangjeong is an important battleground for all who believe in social justice worldwide.

South Korea started construction of the base in January [2011] but protests halted the work in June.

An eyewitness reports that the villagers' nonviolent resistance has led to arrests targeting filmmakers, bloggers, clerics, activists on social-network websites—and most notably, the leaders of the movement.

On August 24, 2011, riot police broke up a nonviolent rally and arrested more than three dozen activists, including the mayor of Gangjeong; the leader of one of the most effective peace groups in Korea; and a Catholic priest.

Basic democratic ideals are also under threat. In the 2007 vote to authorize the construction of the naval base, eighty-seven people, some of whom reportedly were bribed, decided the fate of an entire village of 1,900 and an island of more than a half-million people.

Islanders were told that the military base would double as a tourism hub for cruise ships—indeed, that it would be the only means for such ships to dock at the island, yielding commercial benefits. The claim is hardly credible, if only because at the same time, on a different shore, a massive port expansion project has been under way and could be completed by summer 2012. It has already been announced that this new port will host cruise liners.

Gangjeong villagers know full well what their future holds if their cry for peace is not heeded: an influx of South Korean and foreign military personnel, advanced armaments, and a world of suffering delivered to a small island that has already endured enough. The irony is that the seeds for future superpower conflict are being sown on an ecological preserve and island of peace.

Occupy the Future

OCTOBER 31, 2011

This article is adapted from Noam Chomsky's talk at the Occupy Boston encampment on Dewey Square on October 22, 2011. He spoke as part of the Howard Zinn Memorial Lecture Series held by Occupy Boston's on-site Free University.

Delivering a Howard Zinn lecture is a bittersweet experience for me. I can't help but regret that he's not here to take part in and invigorate a movement that would have been the dream of his life, and for which he laid a lot of the groundwork.

The Occupy movements are exciting, inspiring. If the bonds and associations being established in these remarkable events can be sustained and carried forward through a long, hard period ahead—victories don't come quickly—the Occupy protests could mark a truly significant moment in American history.

I've never seen anything quite like the Occupy movement in scale and character, here and worldwide. The Occupy outposts are trying to create cooperative communities that just might be the basis for the kinds of lasting organizations necessary to overcome the barriers ahead and the backlash that's already coming.

The Occupy movement is in many ways unprecedented. That is natural enough, because this is an unprecedented era, not just at this moment but since the 1970s.

The 1970s marked a turning point for the United States. Since the country began, with ups and downs it had been a developing society, not always in very pretty ways, but with general progress toward industrialization, prosperity and expansion of rights.

Even in dark times, the expectation was that the progress would continue. I'm just old enough to remember the Great Depression. By the mid-1930s, even though the situation was objectively much harsher than today, the spirit was quite different.

A militant labor movement was organizing, the CIO, and workers were even moving on to stage sit-down strikes, which are just one step from taking over the factories and running them themselves, a very frightening development for the business world.

Under popular pressure, New Deal legislation was passed, which didn't end the Depression, but substantially improved the lives of many people. The prevailing sense was that we would get out of the hard times.

Now there's a sense of hopelessness, sometimes despair. During the 1930s, working people could anticipate that the jobs would come back. Today, if you're a worker in manufacturing, with real unemployment practically at Depression levels, you know that those jobs may be gone forever if current policies persist.

That change in popular understanding has evolved since the 1970s, when major changes took place in the social order. One was a sharp reversal as several centuries of industrialization turned to de-industrialization. Of course manufacturing continued, but overseas—very profitable, though harmful to the workforce.

The economy shifted to financialization. Financial institutions expanded enormously. A vicious cycle was set in motion. Wealth concentrated in the financial sector. The cost of campaigns escalated sharply, driving political leaders ever deeper into the pockets of wealthy backers, increasingly in financial institutions.

Naturally, the funders were rewarded by the politicians they put into office, who instituted policies favorable to Wall Street: deregulation, tax changes, relaxation of rules of corporate governance and other measures that intensified the concentration of wealth and carried the vicious cycle forward. The new policies led very quickly to financial crises, unlike earlier years when New Deal legislation was in place and there were none. From the early Reagan years, each crisis has been more serious than the last, leading finally to the latest collapse in 2008. The government once again came to the rescue of Wall Street firms judged to be too big to fail—the implicit government insurance policy that ensures underpricing of risk—with leaders too big to jail.

Today, for the one-tenth of 1 percent of the population who benefited most from these decades of greed and deceit, everything is fine, while for most of the population, real income has almost stagnated or sometimes declined for thirty years.

In 2005, Citigroup—which once again has been saved by government bailout—released a brochure for investors that urged them to put their money into what they called the Plutonomy Index, which identified stocks in companies that cater to the wealthy. The brochure informed investors that the index has greatly outperformed the market ever since the mid-1980s, when the Reagan-Thatcher regime was settling in.

"The world is dividing into two blocs—the plutonomy and the rest," Citigroup summarized. "The U.S., U.K. and Canada are the key plutonomies—economies powered by the wealthy."

As for the non-rich, they're sometimes called the precariat—people who live a precarious existence at the periphery of society. The "periphery," however, has become a substantial proportion of the population in the United States and elsewhere.

So we have the plutonomy and the precariat: the 1 percent and the 99 percent, in the imagery of the Occupy movement—not literal numbers, but the right picture.

The historic reversal in people's confidence about the future is a reflection of tendencies that could become irreversible. The Occupy protests are the first major popular reaction that could change the dynamic.

I've kept to domestic issues. But two dangerous developments in the international arena overshadow everything else.

For the first time in human history, there are real threats to the survival of the human species. Since 1945 we have had nuclear weapons, and it seems a miracle we have survived them. And policies of the Obama administration and its allies are encouraging escalation.

The other threat, of course, is environmental catastrophe. Practically every country in the world is taking at least halting steps to do something about it. The United States is taking steps backward. Large-scale propaganda operations, openly announced by the business community, seek to convince the public that climate change is all a liberal hoax: Why pay attention to these scientists? Congressional Republicans are now dismantling the limited environmental protections put in place by the Nixon administration, a graphic illustration of how power centers have regressed since the 1970s reversal.

If these tendencies persist in the richest, most powerful country in the world, catastrophe won't be averted.

Something must be done in a disciplined, sustained way, and soon. It won't be easy to proceed. There will be hardships and failures—it's inevitable. But unless the process that's taking place here and elsewhere in the country and around the world continues to grow and becomes a major force in society and politics, the chances for a decent future are bleak.

You can't achieve significant initiatives without a large, active, popular base. It's necessary to get out into the country and help people understand what the Occupy movement is about—what they themselves can do, and what the consequences are of not doing anything.

Organizing such a base involves education and activism. Education doesn't mean telling people what to believe—it also means learning from them and with them.

Karl Marx famously said that the task is not just to understand the world but to change it. A variant to keep in mind is that if you want to change the world you'd better try to understand it. That doesn't mean just listening to a talk or reading a book, though that's helpful sometimes. You learn from participating. You learn from others. You learn from the people you're trying to organize. We all have to gain the understanding and the experience to formulate and implement ideas and plans as to how to move forward.

The most exciting aspect of the Occupy movement is the construction of the linkages that are taking place all over. If they can be sustained and expanded, Occupy can lead to dedicated efforts to set society on a more humane course.

Index

Abbas, Kamal, 259
Abbas, Mahmoud, 31, 32, 84, 133, 202
Abkhazia, 101
Abrams, Elliott, 155, 212
The Accidental Empire (Gorenberg), 243
Afghanistan, 53–54
    Durand Line and, 141–142
    Obama and, 102, 120–123
    relationship with Iran, 144
    war in, 221–224
Afpak, 293
Africa, 49, 268
African National Congress, 249
African Union, 45
AIG (American International Group), 197
ALBA, 135–136
al-Jazeera, 253
Allenby, Edmund, 251
Allende, Salvador, 226
al-Qaida, 277
Al-Quds Al-Arabi, 267
American Electric Power, 287
American exceptionalism, 146
American International Group (AIG), 197
American Petroleum Institute, 238
Aminzai, Bakhtar, 121
Amnesty International, 247
Annapolis Conference (2007), 41–44
Anselem, Lewis, 186
Arab League Proposal, 126–127
Arab Peace Initiative, 151–152
Arab Spring, 268, 294
Archer Daniels Midland, 22
Argentina, 247
arms industry, U. S., 241–242
arms race, 38, 298
Ashkenazi, Gabi, 216

assassinations, U. S.-backed, in El Salvador, 178–179
Associated Press, 168
Atlacatl battalion, 178

Bacevich, Andrew, 78, 227
Baghdad
    ethnic cleansing in, 60
    U. S. embassy in, 66, 88, 170, 174
Bahrain, 266
bailouts, 238, 272
Bajaur, Pakistan, 119
Baker, Dean, 115, 196, 287
Baker, Gerald, 271
Baker, Jim, 102
ballistic missile defense (BMD) programs, 39–40, 78–79
Balzar, John, 47
Banco del Sur, 135
Barak, Ehud, 35, 41, 281–282
Al-Barakaat, 48
Barnett, Correlli, 28
Barofsky, Neil, 272
Bartels, Larry M., 109
Battle of Mogadishu, 46–47
Beck, Ulrich, 269
Beinin, Joel, 260
ben-Ami, Shlomo, 104
Benghazi, 267
Benn, Aluf, 127
Bergen, Peter, 52
Berle, A. A., 231–232
Berlin Wall, fall of, 177
Biden, Joe, 113, 173, 201
bin Laden, Osama, 52, 275–279, 291, 293–295
biofuels, 22, 23–24
Blair, Tony, 34
Bolivarian Alternative for Latin America and the Caribbean, 135
Bolivia, 135–138, 167, 247

Boone, Peter, 238
Bouton, Marshall, 58
BP, 87
Branfman, Fred, 256–257
Brazil, 22, 247, 268
Brenner, Robert, 235
Bretton Woods system, 107–108
Britain, 22, 85, 134, 148
Broad, William J., 20
Brookings Institute, 242, 256
The Brothers Karamazov
  (Dostoyevsky), 180–181
Brown, Scott, 191, 192
Brzezinski, Zbigniew, 102
Buergenthal, Thomas, 243
Burke, Jason, 121, 143–144
Bush, George H. W., 77–78, 153,
  180, 202
Bush, George W.
  2008 Palestinian elections and,
    212
  administration of, 19, 64
  Declaration of Principles (2007),
    88–89
  ethanol production and, 22
  North Korea and, 17
  Russia and, 99, 104
  Tony Blair and, 34
  trip to Middle East (2008), 77,
    81–86, 126
Butler, Lee, 73, 95

Cairo, Egypt, 259
Calderón, Felipe, 21
California, 209–210
Canova, Tim, 114
Cantor, Eric, 244
capitalism, state, 106, 271, 273, 288
Cardoso, Fernando, 169
Carey, Alex, 262
Carnegie Endowment, 266
Carothers, Thomas, 178
Carter, Jimmy, 34, 184–185, 249
Cassinga, 249
Caucasus, 102

Ceausescu, Nicolae, 253
Center for Responsive Politics, 112
Central America, 178
Chace, James, 226
Charnobitz, Steve, 270
Chavez, Hugo, 136
Cheney, Richard "Dick," 26, 63
China, 198, 199, 268
  ascendancy of, 79, 142, 195,
    225–229
  human rights in, 299
  Jeju Island and, 298
  world order and, 231–234
Chiozza, Giacomo, 285
Chivers, C. J., 53
CIA, 224
Circincione, Joseph, 68
Citigroup, 303–304
climate change, 304–305
Clinton, Bill
  administration of, 18–19
  Georgia and, 102
  Indonesian invasion of East Timor
    and, 250
  militarization of Mexican border
    by, 23
  NATO's expansion into East
    Germany and, 78, 103
  rejectionism and, 214
  use of military force and, 227
Clinton, Hillary, 64, 116, 203, 204,
  284
Cohen, Roger, 146
Cohen, William, 227
Cold War, 103–104
Cold War II, 37–40
Colombia, 168, 255
Community of Latin American and
  Caribbean States, 231
Congo, 165
Connell-Smith, Gordon, 186
Constable, Pamela, 143
Constitutive Act of the African
  Union, 164
Consumer Protection Agency, 287

containment policy, 38–39
Cook, Steven A., 218
Corfu Channel case, 162
Cornwell, Richard, 45
corporate contributions and elections, 189–190
corporate power, 285–286
corporate rights, 190–191
corporate sector, 261–262
Correa, Max, 23
Correa, Rafael, 167
Crane, David, 265
Crocker, Ryan, 52
Cruickshank, Paul, 52
Cummings, Bruce, 223
Curtis, Mark, 61, 95
Czech Republic, 39, 79

Dahlan, Muhammad, 32, 212
Daily Telegraph (London), 282
Davies, Howard, 265
Dayan, Moshe, 243
Declaration of Principles (2007), 65–66, 88–89
Defense News, 205
defense spending, 174
deficit, U. S., 196, 286–287
democracy, 28, 231, 253, 259–263
Democratic Party, 65–66, 91, 109, 261
deregulation, of financial institutions, 105–106, 235–236, 271
Dewey, John, 108
DeYoung, Karen, 51
Diego Garcia, 215
diplomacy, 117, 122–123
Dominican Republic, 184
Dostoyevsky, 180–181
Dreazen, Yochi, 276
Durand, Henry Mortimer, 141
Durand Line, 141–142

East Jerusalem, 126, 201–206. *see also* Jerusalem
East Timor, 249–250, 268

Eatwell, John, 106
economic strangulation, 80
economy
    Great Moderation and, 271–272
    impact of financial institutions on, 288–289
    Obama and, 111
Edwards, John, 116
Egypt, 219, 255, 266
    Taba, 34–35, 41–42
    uprisings in, 259–260
Eichengreen, Barry, 108
Einstein, Albert, 74
Eisenhower, Dwight D., 270
El Salvador,, 178–179
elections
    2006 Palestinian, 31, 41, 128, 158, 212
    2008 U. S. presidential, 111–112
    2009 Lebanese, 155–157
    cost of, 288, 303
    in Honduras, 185–186
    in Iran, 157
    U. S. 2010 mid-term, 235
Eliav, Aryeh, 94–95
Emanuel, Rahm, 113–114
embassies, 170, 174
environmental catastrophe, threat of, 304
Erlanger, Steven, 171
Erlich, Reese, 157
"Essentials of Post–Cold War Deterrence" (STRATCOM 1995), 73–74
ethanol, 21–22, 24
Ethiopia, 47–48
Etzioni, Amitai, 216
Europe, 39
European Conquest, 135
European Union, 100
exceptionalism, 146, 148
Exxon Mobil, 87

Fadlallah, Sheik Mohammed Hussein, 277

Falk, Richard, 32
Fallon, William, 68
Farr, Warner, 95
Fatah, 32
Fayyad, Salam, 133
Ferguson, Thomas, 112, 190, 237,
  288–289
Fifield, Anna, 20
financial crisis
  changes in the social order and,
    302–303
  Great Moderation and, 271–274
  Obama administration and, 115
  origins of, 235–236
  presidential campaigns and,
    105–109
  world, 269
financial institutions
  corporate propaganda of, 262–263
  deregulation of, 105–106, 235–
    236, 271
  impact of, on economy, 288–289
  Obama and, 196–198
  presidential campaigns (2008),
    112–113
financialization, 235, 288, 303
Foreign Affairs (Rubin and Rashid),
  121
foreign policy and Latin America, 138
The Foreign Policy Disconnect: What
  Americans Want from Our Leaders but
  Don't Get (Page and Bouton), 57–58
France, 148
Fraser, Doug, 261–262
free market system, 198, 271
free trade, 22–24, 135–139, 191, 270
Freedom Flotilla, 211–215, 249,
  283–284
Freeman, Charles, 77
Friedman, Thomas, 155
Funston, Frederick, 99
Fusion Energy Conference (2008), 93

Gaddis, John Lewis, 226–227
Gadhafi, Moammar, 265, 267

Gamsakhurdia, Zviad, 101
Gangjeong, 297–298
Garton Ash, Timothy, 177
Gates, Robert M., 37
Gaviria, César, 169
Gaza, 31, 125–129, 212–213, 249,
  282–283
Gelb, Leslie, 42
Geneva Accord (2003), 43, 281
Geneva Convention (Fourth), 243
Georgia, 99–100, 101
Gerges, Fawaz, 277
Gettleman, Jeffrey, 45, 46, 47
Gheit, Ahmed Aboul, 96
Gillerman, Dan, 281
Ging, John, 128
global warming, 238, 287
Godec, Robert, 255
Golan Heights, 282
Goldman Sachs, 197
Gomory, Ralph, 198
Goossens, Peter, 45
Gorbachev, Mikhail, 77, 102–104,
  177, 180
Gordon, Michael R., 61
Gordon, Philip, 218
Gorenberg, Gershom, 243
Government Executive, 174
Great Game, 118, 123, 141
Great Moderation, 271–272
Greece, 284
Green Line, 42, 44
Greenspan, Alan, 105, 271–272
Group of 77, 76
Group of Eight, 100
Guantánamo, 145
Gunness, Chris, 284
Gush Shalom, 43

Haiti, 159, 184
Halevy, Ephraim, 96
Halimi, Serge, 100
Hamas
  2006 election in, 128
  G. W. Bush and, 84

Israel and, 31–33, 211–213
Obama and, 69–70, 133–134
two-state settlement and, 42–43
Hariri, Saad, 155
Hart-Landsberg, Martin, 198
Hass, Amira, 213
health care system, 115–116, 192–193, 196, 210, 270, 287
Heilbrunn, Jacob, 255
Hersh, Seymour, 97
Hezbollah, 155
high-tech industry, 250
Hodgson, Godfrey, 146–147
Holder, Eric, 276
Honduras, 158–159, 183, 256
Honduras coup (2009), 159, 185–187
Hoodbhoy, Pervez, 119
Hooglund, Eric, 157
Horwirz, Morton, 190
House Appropriations Committee, 287
human rights, 69, 170
in China, 299
under Jimmy Carter, 184–185
violations of, 48, 148, 167–168, 255
Human Rights Watch, 247
humanitarian intervention, 161–166, 283–284

IAEA. *see* International Atomic Energy Agency (IAEA)
IBM, 199
ICJ (International Court of Justice), 161–162, 165, 243
Ickes, Harold, 251
ILO (International Labor Organization), 270–271
India, 85, 93, 172, 195, 268
Indonesia, 250
International Atomic Energy Agency (IAEA), 76, 93, 171–172
International Commission on Intervention and State Sovereignty on Responsibility to Protect (2001), 164

International Committee of the Red Cross, 283
International Court of Justice (ICJ), 161–162, 165, 243
International Labor Organization (ILO), 270–271
International Republican Institute (IRI), 186
International Security Assistance Force (ISAF), 224
intervention, government, 105
Iran, 25–26
Afghanistan relationship with, 144
elections in, 157
Iraq and, 37–38
Non-Proliferation Treaty (NPT), 67, 76
nuclear proliferation in, 37–40, 67–68, 89–90
nuclear weapons and, 27–28, 96–98, 171, 173–174
Obama and, 66–68, 215–219
promotion of democracy in, 28
threat of, 215–219, 233, 242
Iraq, 64–67, 102
Declaration of Principles (2007) and, 65–66
Iran and, 37–38
Obama and, 117–118
oil reserves in, 87–91, 232, 268
Party Line and, 25–27
popular attitudes in, 51–55
responsibility to protect (R2P) and, 165
war in, 25, 57–61
Iraq Petroleum Industry, 87–88
IRI (International Republican Institute), 186
ISAF (International Security Assistance Force), 224
Israel, 69, 172, 201–206
attack on the Freedom Flotilla by, 211–215, 249, 283–284
G. W. Bush's 2008 trip to, 81–86
Gaza and, 31, 125–129, 249

Hamas and, 212–213
naval maneuvers (July 2009), 173
nuclear policy of, 76–77
nuclear weapons and, 94
Palestine and, 31–35, 41–44
rejection of two-state settlement, 243–244
relations between United States and, 31–35, 43–44, 85, 204–206, 242, 250
threat of recognition of Pakistan by the U. N. and, 281–284
two-state international consensus and, 42–43
Israel, attack on Gaza, 249
Israel-Palestine conflict, 75, 151–153, 158, 241–245, 247–251

Jackson, Robert, 52, 276
Japan, 85, 195
Jeju Island, 297–300
Jerusalem
G. W. Bush's 2008 visit to, 83, 85, 126
settlement expansion in, 152, 201–206, 244
jihadi movement, 293–294
John XXIII (Pope), 179
Johnson, Simon, 238
Jones, Clayton, 233
Jones, James, 142
JoongAng Daily, 298
Jordan, 255
Joya, Malalai, 174–175

Kandahar, Afghanistan, 223
Kaplan, Lawrence, 78
Karzai, Hamid, 37, 120–121, 143, 144
Kattan, Victor, 282
Kayani, Parvez, 119
Kennan, George, 49
Kennedy, John F., 58
Kepel, Gilles, 294
Kerry, John, 153, 204
Kessler, Glenn, 244

Keynes, John Maynard, 107
Kfoury, Assaf, 155
Khouri, Rami, 34
King, Stephen, 233
Kohl, Helmut, 180
Kosovo, 248
Kramer, Andrew E,, 87
Krepon, Michael, 93
Krugman, Paul, 146
Kull, Steven, 286
Kung, Hans, 179
Kurtzer, Daniel C., 126
Kuwait, 266

labor, 269–273
Labor Day, 270
labor movements and democracy, 259–263
Laroche, Eric, 46, 47
Latin America, 135–139, 167–170, 183–187
Latin American Commission on Drugs and Democracy, 169
Law Day, 270
Lebanon, 127, 155–157
Lee, Ching Kwan, 208–209
liberation theology, 179, 180
Libya, 265–268
Lieberman, Avigdor, 44
Lieven, Anatol, 278, 291–293
Lippmann, Walter, 57
Llorens, Hugo, 185, 256
Louis, George, 79
Loyalty Day, 270

Madison, Wisconsin, 259
Maguire, Mairead, 158
Mahalla, Egypt, 259–260
al-Maliki, Nouri, 88, 90
Mandela, Nelson, 249
Maoz, Zeev, 76, 94
Margolis, Eric S., 293
Marja, Afghanistan, 222
Marx, Karl, 14, 305
Massachusetts, 192, 269–270

Massachusetts Bay Colony, 147
Matlock, Jack, 102, 103
Matta, Nada, 259–260
May Day, 269
Maynes, Charles William, 47
Mayr, Ernst, 81
McCain, John, 65, 112
McChrystal, Stanley A., 221
McCoy, Alfred, 147
McGlynn, John, 80
McGwire, Michael, 103
McKinney, Cynthia, 158
Merkel, Angela, 177
Meron, Theodor, 243
Merrill, John, 297
Meshal, Khaled, 42–43
Mexico, 21–24
Miliband, David, 99–100
military bases, 167–170
military coups, 22, 159, 185–187
military spending, 196
Mill, John Stuart, 148
Milne, Seamus, 88
missile defense system, 39–40, 78–79
Mitchell, George, 131, 134, 203
Mogadishu, 46–47
Mohamed El-Baradei, 26
Monitor Group, 265
Morales, Evo, 137, 167, 169
Morocco, 268
Morris, Benny, 94, 96–97
Mottaki, Manouchehr, 26
Muasher, Marwan, 254, 256
Mubarak, Hosni, 253–254, 259–260, 262
Mullen, Michael, 216
Munck, Ronaldo, 269
My Lai massacre, 148
The Myth of American Exceptionalism (Hodgson), 146

NAFTA (North American Free Trade Agreement), 23
National Democratic Institute (NDI), 186

National Intelligence Estimate, 68
National Liberation Front (NLF), 222
National Peace Jirga, 121
National Security Council, 138, 167
NATO (North Atlantic Treaty Organization), 77–78, 102–103, 118, 142, 164–165, 228
NDI (National Democratic Institute), 186
Netanyahu, Benjamin, 127, 201–203, 204
New York Times, 185, 287
Nicaragua, 184–185
Nicholson, Larry, 222
Niebuhr, Reinhold, 224
NLF (National Liberation Front), 222
Nobel Peace Prize, 171, 174
no-fly zone, 267
Non-Aligned Movement, 219
Non-Proliferation Treaty (NPT), 19, 93–94, 172, 218
    Conference (May 2010), 219
    Iran and, 67, 76
North American Free Trade Agreement (NAFTA), 23
North Atlantic Treaty Organization. *see* NATO (North Atlantic Treaty Organization)
North Korea, 17–21
Nuclear Non-Proliferation Treaty (NPT). *see* Non-Proliferation Treaty (NPT)
nuclear proliferation, 27, 37–40, 89–90
nuclear weapons, 73–75, 304
    in Iran, 171
    Iran and, 27–28, 95–98, 171, 173–174
    Israel and, 94
nuclear-weapons-free zones (NWFZs), 75–76, 96, 103
Nuremberg Tribunal, 52, 64–65

Obama, Barack
  administration of, 115, 241
  Afghanistan and, 102, 120–123
  cycle of violence and, 117–123
  economy and, 111
  financial institutions and, 196–198
  Hamas and, 69–70, 133–134
  health care system and, 116
  Honduras coup (2009) and, 159, 185–187
  Iran and, 66–68, 215–219
  Iraq and, 117–118
  Israel-Palestine conflict and, 69–70, 125–129, 131–132, 151–153
  meetings with Netanyahu and, 202
  Middle East and, 117
  nuclear weapons and, 95–96
  Pakistan and, 119–120, 170, 278
  peace efforts of, 171–175
  rejectionism, U. S.-Israeli and, 126, 132, 134
  settlement expansion, 203–204
  support of Israel, 127
  West Bank and, 132–133
Occupy protests, 301–306
Ofer, Idan, 281
offshoring, of production, 288
oil reserves
  in Iraq, 87–91, 118, 232, 268
  in Libya, 265–268
Olmert, Ehud, 44, 128
Olsen, Norman, 32, 212
Oppel, Richard A., Jr., 222
Oren, Amir, 43
Organization of American States, 159, 164
Orwell, George, 101, 209–210
Oslo Accords, 283
Ould-Abdallah, Ahmedou, 46, 47
Oxford Research Business, 59

Page, Benjamin, 58
Pakistan
  death of bin Laden and, 275–279

  Durand Line and, 141
  "good news" from, 54
  nuclear weapons in, 172, 173, 218
  Obama and, 119–120, 170, 278
  pipeline development in, 218, 233
  radicalization of, 291–295
  U. S. embassy in, 256–257
Palestine, 39–40, 69
  Israel and, 31–35, 41–44
  recognition of, 202–204, 247–248, 281–284
  wall in, 177–178
Palestine: Peace not Apartheid (Carter), 34
Palestinian Authority, 31, 282
Party Line, 25–26, 43
Pastor, Robert, 184–185
Patriot Act, 80
Patterson, Anne, 293
Pearl Harbor, 59
Pentagon, 61, 68, 172
Peres, Shimon, 126–127
Perino, Dana, 63
Petraeus, David, 52, 60, 119–120, 201–202, 221
Pike, Douglas, 222
Pinochet, Augusto, 226
PIPA (Program on International Policy Attitudes), 64, 67, 286
pipelines, oil and gas, 102, 118, 218, 233
Plesch, Dan, 215
Plutonomy Index, 303–304
Podesta, John, 114
Poland, 39, 79, 103
Postol, Theodore, 79
precariats, 269, 271, 304
presidential campaigns (2008)
  financial crisis and, 105–109
  financial institutions, 112–113
  Iraq war and, 91
  Middle East policy and, 63–70
The Price of Fear (Warde), 48–49
Program on International Policy Attitudes (PIPA), 64, 67, 286

Prosor, Ron, 282
public opinion
    Arab, 256
    in Egypt, 260
    health care legislation and,
        192–193
    U. S. political parties and, 111
    war in Iraq and, 63–64

al-Qaida, 221–222
Quershi, Shah Mehmood, 120

Rachman, Gideon, 255–256
Raddatz, Martha, 63
Rafael, 250
Rashid, Ahmed, 121
Reagan, Ronald, 236, 249, 254, 292
rejectionism, U. S.-Israeli
    Freedom Flotilla and, 283
    G. W. Bush and, 43
    Obama and, 126, 132, 134
    Taba negotiations and, 34–35
Republican Party, 109, 242, 261, 305
responsibility to protect (R2P), 14,
    161–166
Revolutionary Guards, 38
Rice, Condoleeza, 26, 34, 37, 89, 99,
    212
Rice, Susan, 282
Rickards, James G., 105
Road Map of the Quartet, 33, 133,
    152–153, 203
Roberts, John G., 189
Robertson, Geoffrey, 275, 279
Roh Moo-hyun, 297
Romania, 253
Romero, Oscar, 178
Roosevelt, Franklin D., 59
Roosevelt, Theodore, 183
Rose, David, 32, 212
Rosen, Nir, 60, 90
Rosenberg, Carol, 12
Ross, Dennis, 267
Rubin, Barnett, 121
Rubin, Robert, 114–115

Ruigrok, Winfried, 105
Runge, C. Ford, 22, 23
Russell, Bertrand, 74
Russia, 39, 77–78, 100–104, 268
Russia-Georgia-Ossetia war, 99
Ryan, Kevin, 66

Saakashvili, Mikheil, 100, 101
al-Sadr, Moktada, 90
Salas, Carlos, 23
Samson complex, 94–95
Sanger, David E., 20
Saudi Arabia
    arms sale to, 38
    democracy uprisings in, 266
    G W. Bush's 2008 trip to, 81–86
    oil reserves and exports of, 232
    radical Islam and, 254
Saudi Gazette, 85–86
Scheffer, Jaap de Hoop, 118
Scheuer, Michael, 293–294
Schlesinger, Arthur M., Jr., 58–59
Schmidt, Helmut, 276
School of the Americas, 180
Schoultz, Lars, 147–148
Schwarzenegger, Arnold, 209
Scowcroft, Brent, 162
sectarian warfare. in Iraq, 60
Securities and Exchange Commission,
    287
Senate Armed Services Committee
    Report on Detainee Treatment, 145
Senauer, Benjamin, 22, 23
Separation Wall, 83, 152, 244
September 11, 2001, 277, 291–295
settlement expansion, 42–43, 70, 152,
    201–206, 241–245
Shamir, Yitzhak, 202
Shanghai Cooperation Organization,
    142, 232–233
Shankar, Thom, 225
Shapira, Ya'akov Shimson, 243
Sharon, Ariel, 132
Shell, 87
Sigal, Leon V., 18

Simon, Steven, 91
Smith, Adam, 196, 236–237
SOFA (Status of Forces Agreement), 117
Solow, Robert, 289
Somalia, 45–49
South Africa, 248–249, 268
South America, 79, 231
South Korea, 225–226, 297–300
South Ossetia, 101
South Summit (Cuba 2000), 162
Spirit of Humanity, 158
Stack, Joe, 207–208
state capitalism, 107
Status of Forces Agreement (SOFA), 117
Stern, Fritz, 238
Stevens, John Paul, 190
Stiglitz, Joseph, 196
Stimson, Henry L., 183
Stolberg, Sheryl Gay, 171
Suleiman, Omar, 254
Summers, Lawrence, 114–115
Sunday Herald (Glasgow), 215
Supreme Court (U. S.), 189, 256

Taba, Egypt, 34–35, 41–42
Taking the Risk Out of Democracy (Carey), 262
Taliban, 53, 121–122, 222, 278, 292
TARP (Troubled Asset Relief Program), 197
Taylor, Charles, 265
Taylor, Lance, 106
Tea Party movement, 209, 235, 236
Tehran, 90
Thucydides, 161, 165
Tiedeman, Christopher G., 190
tortillas, 21–24
torture, 145–149
Total, 87
Truman, Harry S., 275–276
Tunisia, 253, 254–255, 266
Turkey, 217–218, 233
Twain, Mark, 99

two-state international consensus, 42–43, 69–70, 151, 214, 243–244, 281

U. S. Congress, 287
U. S. Fourth Fleet, 170
U. S. National Intelligence Estimate, 97
UNASUR (Union of South American Nations), 136, 137, 167–169
unemployment, U. S., 192, 271, 286, 302
UNICEF, 165
Union of South American Nations (UNASUR), 136, 137, 167–169
United Nations (U. N.), 46, 99, 161, 282
    Charter, 95
    Committee on Disarmament, 75
    Declaration on Friendly Relations (1970), 162
    General Assembly, 281–284
    Human Development Index, 195
    Panel on Threats, Challenges and Change, 162–163
    Relief and Works Agency, 128
    World Summit (2005), 161, 163
United Nations (U. N.) Security Council, 163, 243–244
    Resolution 242 (1967), 214
    Resolution 687 (1991), 38–39, 76
    Resolution 1725, 47–48
    Resolution 1887 (2009), 171–173
United States
    attack on Osama bin Laden, 275–279
    containment policy of, 37–40
    decline of, 285–289
    deficit, 195–196, 286–287
    embassy in Baghdad, 66, 88, 170, 174
    embassy in Pakistan, 256–257
    Honduras coup (2009) and, 159, 185–187
    Horn of Africa and, 45, 49
    military bases, in Colombia, 168

military dominance and, 225–229
modes of control by, 80
North Korea and, 17–21
recognition of Palestine, 202–204
relations between Israel and, 31–35, 43–44, 85, 204–206, 242, 250
Somalia and, 45–49
South Africa and, 249
Supreme Court of, 189, 256
Universal Declaration of Human Rights, 108, 166
Unpeoplehood, 61, 95–97
UNRWA (United Nations Relief and Works Agency), 128
USS George Washington, 225

van Creveld, Martin, 27
van Tulder, Rob, 105
Vatican II, 179
Venezuela, 136, 159
vetoes, U. N. Security Council, 163
Vietnam, 58, 148, 222–223
violence
cycle of, and Obama, 117–123
as form of control, 80

Walker, Scott, 263
Wall Street Journal, 57, 114
War Logs, 221
"war on drugs," 169–170
"war on terror," 45, 48–49, 169, 178, 249
Warde, Ibrahim, 48–49
Washington Post, 102, 118, 120

waterboarding, 145, 147
Watergate, 148
weapons-of-mass-destruction-free zone (WMDFZ), 39, 76–77
Weil, Jonathan, 115
Weisbrot, Mark, 97, 158
Weiss, Leonard, 173
West Bank, 248, 282. *see also* settlement expansion
Israeli activity in, 32, 34
land-swap policy of Lieberman for, 44
Obama and, 132–133
U. S. support for Israel in, 42–43
Western Sahara, 268
Whitbeck, John, 247–248
Wikileaks, 221–224, 242, 250, 255–256, 293
Wilson, Woodrow, 58, 184, 191
Winthrop, John, 147
World Food Program, 45
World Summit (2005), 161, 163
Wright, Robin, 37

Yusuf, Moeed, 218

Zakaria, Fareed, 100
Zardari, Asif Ali, 119–120
Zedillo, Ernesto, 169
Zelaya, Manuel, 158, 185, 256
Zia-ul-Haq, Muhammad, 254, 292
Zinni, Anthony, 47
Zionism, 250–251
Zunes, Stephen, 138, 255